Math for Scientists

Natasha Maurits • Branislava Ćurčić-Blake

Math for Scientists

Refreshing the Essentials

Natasha Maurits
Department of Neurology
University Medical Center Groningen
Groningen, The Netherlands

Branislava Ćurčić-Blake
Neuroimaging Center
University Medical Center Groningen
Groningen, The Netherlands

ISBN 978-3-319-57353-3   ISBN 978-3-319-57354-0   (eBook)
DOI 10.1007/978-3-319-57354-0

Library of Congress Control Number: 2017943515

This Springer imprint is published by Springer Nature
The registered company is Springer International Publishing AG
The registered company address is: Gewerbestrasse 11, 6330 Cham, Switzerland

Preface

Almost every student or scientist will at some point run into mathematical formulas or ideas in scientific papers that may be hard to understand or apply, given that formal math education may be some years ago. These math issues can range from reading and understanding mathematical symbols and formulas to using complex numbers, dealing with equations involved in calculating medication equivalents, applying the General Linear Model (GLM) used in, e.g., neuroimaging analysis, finding the minimum of a function, applying independent component analysis, or choosing the best filtering approach. In this book we explain the theory behind many of these mathematical ideas and methods and provide readers with the tools to better understand them. We revisit high-school mathematics and extend and relate them to the mathematics you need to understand and apply the math you may encounter in the course of your research. In addition, this book teaches you to understand the math and formulas in the scientific papers you read. To achieve this goal, each chapter mixes theory with practical pen-and-paper exercises so you (re)gain experience by solving math problems yourself. To provide context, clarify the math, and help readers apply it, each chapter contains real-world and scientific examples. We have also aimed to convey an intuitive understanding of many abstract mathematical concepts.

This book was inspired by a lecture series we developed for junior neuroscientists with very diverse scientific backgrounds, ranging from psychology to linguistics. The initial idea for this lecture series was sparked by a PhD student, who surprised Dr. Ćurčić-Blake by not being able to manipulate an equation that involved exponentials, even though she was very bright. Initially, the PhD student even sought help from a statistician who provided a very complex method to calculate the result she was looking for, which she then implemented in the statistical package SPSS. Yet, simple pen-and-paper exponential and logarithm arithmetic would have solved the problem. Asking around in our departments showed that the problem this particular PhD student encountered was just an example of a more widespread problem and it turned out that many more junior (as well as senior) researchers would be interested in a refresher course about the essentials of mathematics. The first run of lectures in 2014 got very positive feedback from the participants, showing that there is a need for mathematics explained in an accessible way for a broad scientific audience and that the authors' approach provided that. Since then, we have used our students' feedback to improve our approach and this book and its affordable paperback format now make this approach to refreshing the 'math you know you knew' accessible for a wide readership.

Instead of developing a completely new course, we could have tried to build our course on an existing introductory mathematics book. And of course there are ample potentially suitable mathematics books around. Yet, we find that most are too difficult when you are just looking for a quick introduction to what you learned in high school but forgot about. In addition, most mathematics books that are aimed at bachelor-and-up students or non-mathematician researchers present mathematics in a mathematical way, with strict rigor, forgetting that readers like to gain an intuitive understanding and ascertain the purpose of what they are learning. Furthermore, many students and researchers who did not study mathematics can have trouble reading and understanding mathematical symbols and equations. Even though our book is not void of mathematical symbols and equations, the introduction to each mathematical topic is more gradual, taking the reader along, so that the actual mathematics becomes more understandable. With our own firm backgrounds in mathematics (Prof. Maurits) and physics (Dr. Ćurčić-Blake) and our working experience and collaborations in the fields of biophysical chemistry, neurology, psychology, computer science, linguistics, biophysics, and neuroscience, we feel that we have the rather unique combination of skills to write this book.

We envisage that undergraduate students and scientists (from PhD students to professors) in disciplines that build on or make use of mathematical principles, such as neuroscience, biology, psychology, or economics, would find this book helpful. The book can be used as a basis for a refresher course of the essentials of (mostly high-school) mathematics, as we use it now. It is also suited for self-study, since we provide ample examples, references, exercises, and solutions. The book can also be used as a reference book, because most chapters can be read and studied independently. In those cases where earlier discussed topics are needed, we refer to them.

We owe gratitude to several people who have helped us in the process of writing this book. First and foremost, we would like to thank the students of our refresher course for their critical but helpful feedback. Because they did many exercises in the book first, they also helped us to correct errors in answers. The course we developed was also partially taught by other scientists who helped us shape the book and kindly provided some materials. Thank you Dr. Cris Lanting, Dr. Jan Bernard Marsman, and Dr. Remco Renken. Professor Arthur Veldman critically proofread several chapters, which helped incredibly in, especially, clarifying some (too) complicated examples.

Dr. Ćurčić-Blake thanks her math school teachers from Tuzla, whom she appreciates and always had a good understanding with. While the high-school math was very easy, she had to put some very hard work in to grasp the math that was taught in her studies of physics. This is why she highly values Professor Milan Vujičić (who taught mathematical physics) and Professor Enes Udovičić (who taught mathematics 1 and 2) from Belgrade University who encouraged her to do her best and to learn math. She would like to thank her colleagues for giving her ideas for the book and Prof. Maurits for doing the majority of work for this book. Her personal thanks go to her parents Branislav and Spasenka, who always supported her,

her sons Danilo and Matea for being happy children, and her husband Graeme Blake for enabling her, while writing chapters of this book.

One of the professional tasks Professor Maurits enjoys most is teaching and supervising master students and PhD students, finding it very inspiring to see sparks of understanding and inspiration ignite in these junior scientists. With this book she hopes to ignite a similar spark of understanding and hopefully enjoyment toward mathematics in a wide audience of scientists, similar to how the many math teachers she has had since high school did in her. She thanks her students for asking math questions that had her dive into the basics of mathematics again and appreciate it once more for its logic and beauty, her parents for supporting her to study mathematics and become the person and researcher she is now, and, last but not least, Johan for bearing with her through the writing of 'yet' another book and providing many cups of tea.

Finally, we thank you, the reader, for opening this book in an effort to gain more understanding of mathematics. We hope you enjoy reading it, that it gives you answers to your questions, and that it may help you in your scientific endeavors.

Groningen, The Netherlands   Natasha Maurits
Groningen, The Netherlands   Branislava Ćurčić-Blake
April 2017

Contents

Abbreviations

2D	Two-dimensional
3D	Three-dimensional
adj	Adjoint
BEDMAS	(mnemonic) brackets-exponent-division-multiplication-addition-subtraction
BOLD	Blood oxygen level dependent
BP	Blood pressure
cos	Cosine
cov	Covariance
DCM	Dynamic causal modeling
det	Determinant
DFT	Discrete Fourier transform
DTI	Diffusion tensor imaging
EEG	Electroencephalography
EMG	Electromyography
FFT	Fast Fourier transform
(f)MRI	(functional) Magnetic resonance imaging
GLM	General linear model
HRF	Hemodynamic response function
ICA	Independent component analysis
iff	If and only if
lim	Limit
PCA	Principal component analysis
PEMA	(mnemonic) parentheses-exponent-multiplication-addition
sin	Sine
SOHCAHTOA	(mnemonic) Sine (opposite over hypotenuse) Cosine (adjacent over hypotenuse) Tangent (opposite over adjacent)
SVD	Singular value decomposition
tan	Tangent

1

Numbers and Mathematical Symbols

Natasha Maurits

After reading this chapter you know:

- what numbers are and why they are used,
- what number classes are and how they are related to each other,
- what numeral systems are,
- the metric prefixes,
- how to do arithmetic with fractions,
- what complex numbers are, how they can be represented and how to do arithmetic with them,
- the most common mathematical symbols and
- how to get an understanding of mathematical formulas.

1.1 What Are Numbers and Mathematical Symbols and Why Are They Used?

A refresher course on mathematics can not start without an introduction to numbers. Firstly, because one of the first study topics for mathematicians were numbers and secondly, because mathematics becomes really hard without a thorough understanding of numbers. The branch of mathematics that studies numbers is called number theory and *arithmetic* forms a part of that. We have all learned arithmetic starting from kindergarten throughout primary school and beyond. This suggests that an introduction to numbers is not even necessary; we use numbers on a day-to-day basis when we count and measure and you might think that numbers hold no mysteries for you. Yet, arithmetic can be as difficult to learn as reading and some people never master it, leading to *dyscalculia*.

N. Maurits (✉)

Department of Neurology, University Medical Center Groningen, Groningen, The Netherlands

e-mail: n.m.maurits@umcg.nl

© Springer International Publishing AG 2017

N. Maurits, B. Ćurčić-Blake, *Math for Scientists*, DOI 10.1007/978-3-319-57354-0_1

So, what is a number? You might say: 'well, five is a number and 243, as well as 1963443295765'. This is all true, but what is the essence of a number? You can think of a number as an abstract representation of a quantity that we can use to measure and count. It is represented by a symbol or numeral, e.g., the number five can be represented by the Arabic numeral 5, by the Roman numeral V, by five fingers, by five dots on a dice, by ⊪, by five abstract symbols such as ••••• and in many other different ways. *Synesthetes* even associate numbers with colors. But, importantly, independent of how a number is represented, the abstract notion of this number does not change.

Most likely, (abstract) numbers were introduced after people had developed a need to count. Counting can be done without numbers, by using fingers, sticks or pebbles to represent single or groups of objects. It allows keeping track of stock and simple communication, but when quantities become larger, this becomes more difficult, even when abstract words for small quantities are available. A more compact way of counting is to put a mark—like a scratch or a line—on a stick or a rock for each counted object. We still use this approach when tally marking. However, marking does not allow dealing with large numbers either. Also, these methods do not allow dealing with negative numbers (as e.g., encountered as debts in accounting), fractions (to indicate a part of a whole) or other even more complex types of numbers.

The reason that we can deal with these more abstract types of numbers, that no longer relate to countable series of objects, is that numeral systems have developed over centuries. In a numeral system a systematic method is used to create number words, so that it is not necessary to remember separate words for all numbers, which would be sheer impossible. Depending on the base that is used, this systematic system differs between languages and cultures. In many current languages and cultures base 10 is used for the numeral system, probably as a result of initially using the 10 digits (fingers and thumbs) to count. In this system, enumeration and numbering is done by tens, hundreds, thousands etcetera. But remnants of older counting systems are still visible, e.g. in the words twelve (which is not ten-two) or quatre-vingts (80 in French; four twenties). For a very interesting, easy to read and thorough treatise on numbers please see Posamenter and Thaller (2015).

We now introduce the first mathematical symbols in this book; for numbers. In the base 10 numeral system the Arabic numerals 0, 1, 2, 3, 4, 5, 6, 7, 8 and 9 are used. In general, mathematical symbols are useful because they help communicating about abstract mathematical structures, and allow presenting such structures in a concise way. In addition, the use of symbols speeds up doing mathematics and communicating about it considerably, also because every symbol in its context only has one single meaning. Interestingly, mathematical symbols do not differ between languages and thus provide a universal language of mathematics. For non-mathematicians, the abstract symbols can pose a problem though, because it is not easy to remember their meaning if they are not used on a daily basis. Later in this chapter, we will therefore introduce and explain often used mathematical symbols and some conventions in writing mathematics. In this and the next chapters, we will also introduce symbols that are specific to the topic discussed in each chapter. They will be summarized at the end of each chapter.

1.2 Classes of Numbers

When you learn to count, you often do so by enumerating a set of objects. There are numerous children (picture) books aiding in this process by showing one ball, two socks, three dolls, four cars etcetera. The first numbers we encounter are thus 1, 2, 3, . . . Note that '. . .' is a mathematical symbol that indicates that the pattern continues. Next comes zero. This is a rather peculiar number, because it is a number that signifies the absence of something. It also has its own few rules regarding arithmetic:

$$a + 0 = a$$
$$a \times 0 = 0$$
$$\frac{a}{0} = \infty$$

Here, a is any number and ∞ is the symbol for infinity, the number that is larger than any countable number.

Together, 0, 1, 2, 3, . . . are referred to as the *natural numbers* with the symbol $\mathbb{N}$. A special class of natural numbers is formed by the *prime numbers* or primes; natural numbers >1 that only have 1 and themselves as positive divisors. The first prime numbers are 2, 3, 5, 7, 11, 13, 17, 19 etcetera. An important application of prime numbers is in *cryptography*, where they make use of the fact that it is very difficult to factor very large numbers into their primes. Because of their use for cryptography and because prime numbers become rarer as numbers get larger, special computer algorithms are nowadays used to find previously unknown primes.

The basis set of natural numbers can be extended to include negative numbers: . . ., −3, −2, −1, 0, 1, 2, 3, . . . Negative numbers arise when larger numbers are subtracted from smaller numbers, as happens e.g. in accounting, or when indicating freezing temperatures indicated in °C (degrees Centigrade). These numbers are referred to as the integer numbers with symbol $\mathbb{Z}$ (for 'zahl', the German word for number). Thus $\mathbb{N}$ is a subset of $\mathbb{Z}$.

By dividing integer numbers by each other or taking their ratio, we get fractions or rational numbers, which are symbolized by $\mathbb{Q}$ (for quotient). Any rational number can be written as a fraction, i.e. a ratio of an integer, the *numerator*, and a positive integer, the *denominator*. As any integer can be written as a fraction, namely the integer itself divided by 1, $\mathbb{Z}$ is a subset of $\mathbb{Q}$. Arithmetic with fractions is difficult to learn for many; to refresh your memory the main rules are therefore repeated in Sect. 1.2.1.

Numbers that can be measured but that can not (always) be expressed as fractions are referred to as *real numbers* with the symbol $\mathbb{R}$. Real numbers are typically represented by decimal numbers, in which the decimal point separates the 'ones' digit from the 'tenths' digit (see also Sect. 1.2.3 on numeral systems) as in 4.23 which is equal to $\frac{423}{100}$. There are *finite decimal numbers* and *infinite decimal numbers*. The latter are often indicated by providing a finite number of the decimals and then the '. . .' symbol to indicate that the sequence continues. For example, $\pi = 3.1415 \ldots$ Real numbers such as π that are not rational are called *irrational*. Any rational number is real, however, and therefore $\mathbb{Q}$ is a subset of $\mathbb{R}$.

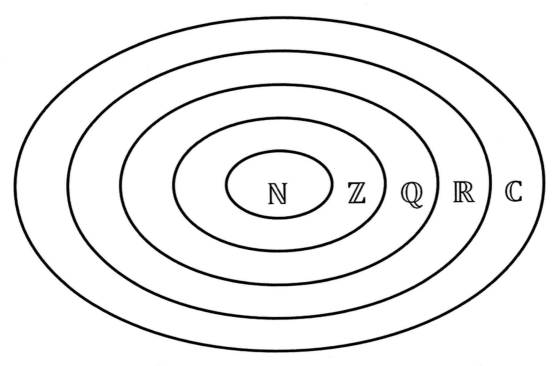

Fig. 1.1 The relationship between the different classes of numbers: $\mathbb{N} \subset \mathbb{Z} \subset \mathbb{Q} \subset \mathbb{R} \subset \mathbb{C}$, where $\subset$ is the symbol for 'is a subset of'.

The last extension of the number sets to be discussed here is the set of *complex numbers* indicated by $\mathbb{C}$. Complex numbers were invented to have solutions for equations such as $x^2 + 1 = 0$. The solution to this equation was defined to be $x = i$. As the complex numbers are abstract, no longer measurable quantities that have their own arithmetic rules and are very useful in scientific applications, they deserve their own section and are discussed in Sect. 1.2.4.

The relationship between the different classes of numbers is summarized in Fig. 1.1.

Exercise

1.1. What is the smallest class of numbers that the following numbers belong to?

a) -7
b) e (Euler's number, approximately equal to 2.71828)
c) $\sqrt{3}$
d) 0.342
e) 543725
f) π
g) $\sqrt{-3}$

1.2.1 Arithmetic with Fractions

For many, there is something confusing about fractional arithmetic, which is the reason we spend a section on explaining it. To add or subtract fractions with unlike denominators you first need to know how to find the smallest *common denominator*. This is the least common multiple, i.e. the smallest number that can be divided by both denominators. Let's illustrate this by some examples.

Suppose you want to add $\frac{2}{3}$ and $\frac{4}{9}$. The denominators are thus 3 and 9. Here, the common denominator is simply 9, because it is the smallest number divisible by both 3 and 9. Thus, if one denominator is divisible by the other, the largest denominator is the common denominator. Let's make it a bit more difficult. When adding $\frac{1}{3}$ and $\frac{3}{4}$ the common denominator is 12; the product of 3 and 4. There is no smaller number that is divisible by both 3 and 4. Note that to find a common denominator, you can <u>always</u> multiply the two denominators. However, this will not always give you the <u>least</u> common multiple and may make working with the fractions unnecessarily complicated. For example, $\frac{7}{9}$ and $\frac{5}{12}$ do have $9 \times 12 = 108$ as a common denominator, but the least common multiple is actually 36. So, how do you find this least common multiple? The straightforward way that always works is to take one of the denominators and look at its table of multiplication. Take the one you know the table of best. For 9 and 12, start looking at the multiples of 9 until you have found a number that is also divisible by 12. Thus, try $2 \times 9 = 18$ (not divisible by 12), $3 \times 9 = 27$ (not divisible by 12) and $4 \times 9 = 36$ (yes, divisible by 12!). Hence, 36 is the least common multiple of 9 and 12. Once you have found a common denominator, you have to rewrite both fractions such that they get this denominator by multiplying the numerator with the same number you needed to multiply the denominator with to get the common denominator. Then you can do the addition. Let's work this out for $\frac{7}{9} + \frac{5}{12}$:

$$\frac{7}{9} + \frac{5}{12} = \frac{7 \times 4}{9 \times 4} + \frac{5 \times 3}{12 \times 3} = \frac{28}{36} + \frac{15}{36} = \frac{43}{36} = 1\frac{7}{36}$$

Note that we have here made use of an important rule for fraction manipulation: whatever number you multiply the denominator with (positive, negative or fractional itself), you also have to multiply the numerator with and vice versa! Subtracting fractions takes the exact same preparatory procedure of finding a common denominator. And adding or subtracting more than two fractions also works the same way; you just have to find a common multiple for all denominators. There is also an unmentioned rule to always provide the simplest form of the result of arithmetic with fractions. The simplest form is obtained by 1) taking out the wholes and then 2) simplifying the resulting fraction by dividing both numerator and denominator by common factors.

Exercise

1.2. Simplify

a) $\dfrac{24}{21}$

b) $\dfrac{60}{48}$

c) $\dfrac{20}{7}$

d) $\dfrac{20}{6}$

1.3. Find the answer (and simplify whenever possible)

a) $\dfrac{1}{3} + \dfrac{2}{5}$

b) $\dfrac{3}{14} + \dfrac{7}{28}$

c) $\dfrac{1}{2} + \dfrac{1}{3} + \dfrac{1}{6}$

d) $\dfrac{3}{4} + \dfrac{7}{8} + \dfrac{9}{20}$

e) $\dfrac{1}{4} - \dfrac{5}{6} + \dfrac{3}{8}$

f) $-\dfrac{1}{3} + \dfrac{1}{6} - \dfrac{1}{7}$

For multiplying and dividing fractions there are two important rules to remember:

1) when multiplying fractions the numerators have to be multiplied to find the new numerator and the denominators have to be multiplied to find the new denominator:

$$\frac{a}{b} \times \frac{c}{d} = \frac{ac}{bd}$$

2) dividing by a fraction is the same as multiplying by the inverse:

$$\frac{a}{b} \div \frac{c}{d} = \frac{a}{b} \times \frac{d}{c} = \frac{ad}{bc}$$

For the latter, we actually make use of the rule that when multiplying the numerator/denominator with a number, the denominator/numerator has to be multiplied with the same number and that to get rid of the fraction in the denominator we have to multiply it by its inverse:

$$\frac{a}{b} \div \frac{c}{d} = \frac{\frac{a}{b}}{\frac{c}{d}} = \frac{\frac{a}{b} \times \frac{d}{c}}{\frac{c}{d} \times \frac{d}{c}} = \frac{\frac{a}{b} \times \frac{d}{c}}{1} = \frac{a}{b} \times \frac{d}{c}$$

Exercise

1.4. Find the answer (and simplify whenever possible)

a) $\dfrac{2}{3} \times \dfrac{6}{7}$

b) $1\dfrac{2}{5} \times 1\dfrac{3}{7}$

c) $\dfrac{5}{6} \div \dfrac{6}{5}$

d) $\dfrac{11}{13} \times \dfrac{2}{3} \div \dfrac{6}{13}$

e) $\dfrac{2}{4} \div 2 \times \dfrac{12}{48}$

Finally, for this section, it is important to note that arithmetic operations have to be applied in a certain order, because the answer depends on this order. For example, when $3 + 4 \times 2$ is seen as $(3 + 4) \times 2$ the answer is 14, whereas when it is seen as $3 + (4 \times 2)$ the answer is 11. The order of arithmetic operations is the following:

1) brackets (or parentheses)
2) exponents and roots
3) multiplication and division
4) addition and subtraction

There are several mnemonics around to remember this order, such as BEDMAS, which stands for Brackets-Exponent-Division-Multiplication-Addition-Subtraction. The simplest mnemonic is PEMA for Parentheses-Exponent-Multiplication-Addition; it assumes that you know that exponents and roots are at the same level, as are multiplication and division and addition and subtraction. Think a little bit about this. When you realize that subtracting a number is the same as adding a negative number, dividing by a number is the same as multiplying by its inverse and taking the nth root is the same as raising the number to the power $1/n$, this makes perfect sense (see also Sect. 1.2.2).

Exercise

1.5. Calculate the answer to

a) $8 \div 4 - 1 \times 3^2 + 3 \times 4$

b) $(8 \div 4 - 1) \times 3^2 + 3 \times 4$

c) $(8 \div 4 - 1) \times (3^2 + 3) \times 4$

d) $(8 \div 4 - 1) \times (3^2 + 3 \times 4)$

1.2.2 Arithmetic with Exponents and Logarithms

Other topics in arithmetic that often cause people problems are exponentials and their inverse, logarithms. Exponentiation is a mathematical operation in which a *base a* is multiplied by itself *n* times and is expressed as:

$$a^n = a \times \cdots \times a$$

Here, *n* is referred to as the *exponent*. Exponentiation is encountered often in daily life, such as in models for population growth or calculations of compound interest. For example, when you have a savings account that yields 2% interest per year, your starting capital of €100,00 will have increased to $100 + \frac{2}{100} \cdot 100 = 1,02 \cdot 100 = 102$. Another year later, you will have $102 + \frac{2}{100} \cdot 102 = 1,02 \cdot 102 = 1,02 \cdot 1,02 \cdot 100 = (1,02)^2 \cdot 100 = 104,04$. Thus, after *n* years, your capital will have grown to $(1,02)^n \cdot 100$. In general, when your bank gives you *p*% interest per year, your starting capital of *C* will have increased to $\left(1 + \frac{p}{100}\right)^n \cdot C$ after *n* years. Here, we clearly see exponentiation at work. Let me here remind you of some arithmetic rules for exponentiation that will come in very handy when continuing with the next chapters (*a* and *b* should be non-zero):

$$\begin{aligned}
a^0 &= 1 \\
a^{-n} &= \frac{1}{a^n} \\
a^n a^m &= a^{n+m} \\
\frac{a^n}{a^m} &= a^{n-m} \\
(a^n)^m &= a^{nm} \\
(ab)^n &= a^n b^n
\end{aligned} \qquad (1.1)$$

Exercise

1.6. Simplify to one power of 2:

  a) $\dfrac{2^3 2^4}{2^2}$

  b) $\dfrac{(2^2)^{\frac{1}{2}} 2^3}{2^{-4} 2^2}$

This is also the perfect point to relate *roots* to exponentials, because it makes arithmetic with roots so much easier. Mostly, when people think of roots, they think of the square root, designated with the symbol $\sqrt{\ }$. A square root of a number *x* is the number *y* such that $y^2 = x$. For example, the square roots of 16 are 4 and -4, because both 4^2 and $(-4)^2$ are 16. More generally, the *n*th root of a number *x* is the number *y* such that $y^n = x$. An example is given by the cube root of 8 which is 2 ($2^3 = 8$). The symbol used for the *n*th root is $\sqrt[n]{\ }$, as in $\sqrt[3]{8} = 2$.

And here comes the link between roots and exponents: $\sqrt[n]{x} = x^{\frac{1}{n}}$. Knowing this relationship, and all arithmetic rules for exponentiation (Eq. 1.1), allows for easy manipulation of roots. For example,

$$\frac{\sqrt[4]{9}}{\sqrt[8]{3}} = \frac{9^{\frac{1}{4}}}{3^{\frac{1}{8}}} = 9^{\frac{1}{4}} \cdot 3^{-\frac{1}{8}} = \left(3^2\right)^{\frac{1}{4}} \cdot 3^{-\frac{1}{8}} = 3^{\frac{1}{2}} \cdot 3^{-\frac{1}{8}} = 3^{\frac{1}{2}-\frac{1}{8}} = 3^{\frac{3}{8}} = \left(3^3\right)^{\frac{1}{8}} = \sqrt[8]{27}$$

Exercise

1.7. Simplify as much as possible:

a) $\dfrac{\sqrt[3]{1000}}{\sqrt[4]{16}}$

b) $\sqrt[4]{25}\sqrt{5}$

c) $\sqrt{3y^8}$

d) $\dfrac{\sqrt[4]{9}}{\sqrt[8]{3}}$ (this is the same fraction as in the example above; now try to simplify by rewriting the fourth root to an eighth root right away)

e) $\sqrt[3]{x^{15}}$

f) $\sqrt[7]{p^{49}}$

g) $\sqrt[3]{\dfrac{a^6}{b^{27}}}$

h) $\sqrt[3]{\dfrac{-27x^6y^9}{64}}$

Finally, I will briefly review the arithmetics of the *logarithm*, the inverse operation of exponentiation. The base n logarithm of a number y is the exponent to which n must be raised to produce y; i.e. $\log_n y = x$ when $n^x = y$. Thus, for example, $\log_{10} 1000 = 3$, $\log_2 16 = 4$ and $\log_7 49 = 2$. A special logarithm is the *natural logarithm*, with base e, referred to as *ln*. The number e has a special status in mathematics, just like π, and is encountered in many applications (see e.g., Sect. 3.2.1). It also naturally arises when calculating compound interest, as it is equal to $\left(1 + \frac{1}{n}\right)^n$ when n goes to infinity (see Sect. 6.7; verify that this expression gets close to e for a few increasing values for n). The basic arithmetic rules for logarithms are:

$$\log_b y^a = a\log_b y$$

$$\log_b \sqrt[a]{y} = \frac{\log_b y}{a}$$

$$\log_b xy = \log_b x + \log_b y$$

$$\log_b \frac{x}{y} = \log_b x - \log_b y$$

$$\log_b y = \frac{\log_k y}{\log_k b}$$

The third arithmetic rule above shows that logarithms turn multiplication into addition, which is generally much easier. This was the reason that, before the age of calculators and computers (until approximately 1980), logarithms were used to simplify the multiplication of large numbers by means of slide rules and logarithm tables.

Exercise

1.8. Simplify as much as possible:

a) $\dfrac{\log_b\left(\left[x^2+1\right]^4\right)}{\log_b\sqrt{x}}$

b) $\log_2(8\cdot 2^x)$

c) $\dfrac{1}{\log_{27}(3)}$

d) $\log_2(8\cdot\sqrt[3]{8})$

1.9. Rewrite to one logarithm:

a) $\log_2 x^2 + \log_2 5 + \log_2\frac{1}{3}$

b) $\log_3\sqrt{a} + \log_3(10) - \log_3 a^2$

c) $\log_a a^2 - \log_a 3 + \log_a\frac{1}{3}$

d) $\log_x\sqrt{x} + \log_x x^2 + \log_x\frac{1}{\sqrt{x}}$

1.2.3 Numeral Systems

In the Roman numeral system, the value of a numeral is independent of its position: I is always 1, V is always 5, X is always 10 and C is always 100, although the value of the numeral has to be subtracted from the next numeral if that is larger (e.g., IV = 4 and XL = 40). Hence, XXV is 25 and CXXIV is 124. This way of noting numbers becomes inconvenient for larger numbers (e.g., 858 in Roman numerals is DCCCLVIII, although because of the subtraction rule 958 in Roman numerals is CMLVIII). In the most widely used numeral system today, the decimal system, the value of a numeral does depend on its position. For example, the 1 in 1 means one, while it means ten in 12 and 100 in 175. Such a positional or place-value notation allows for a very compact way of denoting numbers in which only as many symbols are needed as the base size, i.e., 10 (0,1,2,3,4,5,6,7,8,9) for the decimal system which is a base-10 system. Furthermore, arithmetic in the decimal system is much easier than in the Roman numeral system. You are probably pleased not to be a Roman child having to do additions! The now commonly used base-10 system probably is a heritage of using ten fingers to count. Since not all counting systems used ten fingers, but also e.g., the three phalanges of the four fingers on one hand or the ten fingers and ten toes, other numerical bases have also been around for a long time and several are still in use today, such as the duodecimal or base-12 system for counting hours and months. In addition, new numerical bases have been introduced because of their convenience for certain specific purposes, like the binary (base-2) system for digital computing. To understand which number numerals indicate, it is important to know the base that is used, e.g. 11 means eleven in the decimal system but 3 in the binary system.

To understand the systematics of the different numeral systems it is important to realize that the position of the numeral indicates the power of the base it has to be multiplied with to give its value. This may sound complicated, so let's work it out for some examples in the decimal system first:

$$154 = 1 \times 10^2 + 5 \times 10^1 + 4 \times 10^0$$
$$= 1 \times 100 + 5 \times 10 + 4 \times 1$$
$$3076 = 3 \times 10^3 + 0 \times 10^2 + 7 \times 10^1 + 6 \times 10^0$$
$$= 3 \times 1000 + 0 \times 100 + 7 \times 10 + 6 \times 1$$

Hence, from right to left, the power of the base increases from $base^0$ to $base^1$, $base^2$ etcetera. Note that the 0 numeral is very important here, because it indicates that a power is absent in the number. This concept works just the same for binary systems, only the base is different and just two digits, 0 and 1 are used:

$$101 = 1 \times 2^2 + 0 \times 2^1 + 1 \times 2^0$$
$$= 1 \times 4 + 0 \times 2 + 1 \times 1$$
$$110011 = 1 \times 2^5 + 1 \times 2^4 + 0 \times 2^3 + 0 \times 2^2 + 1 \times 2^1 + 1 \times 2^0$$
$$= 1 \times 32 + 1 \times 16 + 0 \times 8 + 0 \times 4 + 1 \times 2 + 1 \times 1$$

Thus 101 and 110011 in the binary system are equal to 5 and 51 in the decimal system. An overview of these two numeral systems is provided in Table 1.1.

In the binary system a one-positional number is indicated as a *bit* and an eight-positional number (consisting of 8 bits) is indicated as a *byte*.

Exercise

1.10. Convert these binary numbers to their decimal counterparts
  a) 10
  b) 111
  c) 1011
  d) 10101
  e) 111111
  f) 1001001

Table 1.1 The first seven powers used for the place values in the decimal (base 10) and the binary (base 2) systems

Power	7	6	5	4	3	2	1	0
Value in decimal system	10,000,000	1,000,000	100,000	10,000	1000	100	10	1
Value in binary system	128	64	32	16	8	4	2	1

There are some special notations for numbers in the decimal system, that are easy to know and use. To easily handle very large or very small numbers with few *significant digits*, i.e., numbers with many trailing or leading zeroes, the *scientific notation* is used in which the insignificant zeroes are more or less replaced by their related power of 10. Consider these examples:

$$10000 = 1 \times 10^4$$

$$0.0001 = 1 \times 10^{-4}$$

$$5340000 = 5.34 \times 10^6$$

$$0.00372 = 3.72 \times 10^{-3}$$

$$696352000000000 = 6.96352 \times 10^{14}$$

To get the scientific notation of a number, you thus have to count the number of digits the comma has to be shifted to the right (positive powers) or to the left (negative powers) to arrive at the original representation of the number. Calculators will use 'E' instead of the 10 base, e.g., $10000 = 1E4$.

Exercise

1.11. Write in scientific notation

   a) 54000
   b) 0.0036
   c) 100
   d) 0.00001
   e) 654300
   f) 0.000000000742

To finalize this section on numeral systems I would like to remind you of the metric prefixes, that are used to indicate a multiple or a fraction of a unit and precede the unit. This may sound cryptic, but what I mean are the 'milli-' in millimeter and the 'kilo-' in kilogram, for example. The reason to introduce them here is that nowadays, all metric prefixes are related to the decimal system. Table 1.2 presents the most commonly used prefixes.

1.2.4 Complex Numbers

In general, complex numbers extend the one-dimensional world of real numbers to two dimensions by including a second, *imaginary* number. The complex number i, which indicates the imaginary unit, is defined as the (positive) solution to the equation $x^2 + 1 = 0$, or, in other words, i is the square root of -1. Every complex number is characterized by a pair of numbers (a,b), where a is the real part and b the imaginary part of the number. In this sense a complex number can also be seen geometrically as a *vector* (see also Chap. 4) in the complex plane (see Fig. 1.2). This complex plane is a 2-dimensional coordinate system where

Table 1.2 The most commonly used metric prefixes, their symbols, associated multiplication factors and powers of 10

Prefix	Symbol	Factor	Power of 10
Exa	E	1000 000 000 000 000 000	18
Peta	P	1000 000 000 000 000	15
Tera	T	1000 000 000 000	12
Giga	G	1000 000 000	9
Mega	M	1000 000	6
Kilo	k	1000	3
Hecto	h	100	2
Deca	da	10	1
Deci	d	0.1	−1
Centi	c	0.01	−2
Milli	m	0.001	−3
Micro	μ	0.000 001	−6
Nano	n	0.000 000 001	−9
Pico	p	0.000 000 000 001	−12
Femto	f	0.000 000 000 000 001	−15

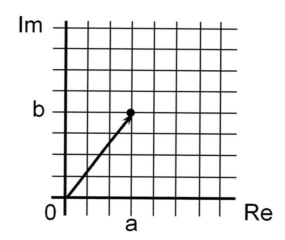

Fig. 1.2 Illustration of the complex number $a + bi$ as a pair or vector in the complex plane. *Re* real axis, *Im* imaginary axis.

the real part of the complex number indicates the distance to the vertical *axis* (or reference line) and the imaginary part of the complex number indicates the distance to the horizontal axis. Both axes meet in the origin. The horizontal axis is also referred to as the real axis and the vertical axis as the imaginary axis. A complex number z is also written as $z = a + bi$. For manipulating complex numbers and working with them, it helps to remember that a complex number has these two representations, i.e. as a pair or vector (a,b) in the two-dimensional complex plane and as a number $a + bi$.

14   N. Maurits

Exercise

1.12. Draw/position the following complex numbers in the complex plane

    a) $1 + i$
    b) $-2 - 2.5i$
    c) $-3 + 2i$
    d) $4\sqrt{-1}$

1.2.4.1 Arithmetic with Complex Numbers

Let's start simple, by adding complex numbers. This is done by adding the real and imaginary parts separately:

$$(a + bi) + (c + di) = (a + c) + (b + d)i$$

Similarly, subtracting two complex numbers is done by subtracting the real and imaginary parts separately:

$$(a + bi) - (c + di) = (a - c) + (b - d)i$$

Alternatively, adding or subtracting two complex numbers can be viewed of geometrically as adding or subtracting the associated vectors in the complex plane by constructing a parallelogram (see Fig. 1.3).

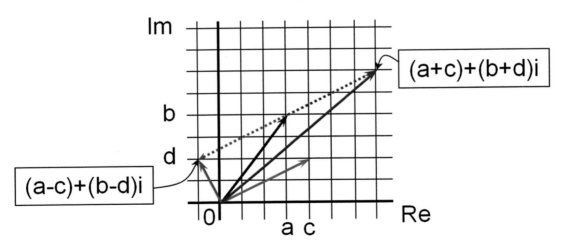

Fig. 1.3 Illustration of adding (*blue*) and subtracting (*green*) the complex numbers $a + bi$ (*black*) and $c + di$ (*red*) in the complex plane. The *dashed arrows* indicate how $c + di$ is added to (*blue dashed*) or subtracted from (*green dashed*) $a + bi$.

Multiplying two complex numbers is done by using the *distributive law* (multiplying the two elements of the first complex number with each of the two elements of the second complex number and adding them):

$$(a + bi)(c + di) = ac + adi + bci + bdi^2 = (ac - bd) + (ad + bc)i \qquad (1.2)$$

Here, we make use of the fact that $i^2 = -1$. Finally, division of two complex numbers is done by first multiplying numerator and denominator by the *complex conjugate* of the denominator (and then applying the distributive law again) to make the denominator real:

$$\frac{a + bi}{c + di} = \frac{(a + bi)\,(c - di)}{(c + di)\,(c - di)} = \frac{ac - adi + bci - bdi^2}{c^2 - cdi + cdi - d^2i^2}$$
$$= \frac{(ac + bd) + (bc - ad)i}{c^2 + d^2} = \frac{ac + bd}{c^2 + d^2} + \frac{bc - ad}{c^2 + d^2}i$$

The complex conjugate of a complex number is indicated by an overbar and is calculated as:

$$\overline{a + bi} = a - bi$$

Hence, for a complex number $z = a + bi$:

$$z\bar{z} = a^2 + b^2$$

Exercise

1.13. Calculate:

    a) $(1 + i) + (-2 + 3i)$
    b) $(1.1 - 3.7i) + (-0.6 + 2.2i)$
    c) $(2 + 3i) - (2 - 5i)$
    d) $(4 - 6i) - (6 + 4i)$
    e) $(2 + 2i) \times (3 - 3i)$
    f) $(5 - 4i) \times (1 - i)$
    g) $\dfrac{5 - 10i}{1 - 2i}$
    h) $\dfrac{18 + 9i}{\sqrt{5} - 2i}$

1.2.4.2 The Polar Form of Complex Numbers

An alternative and often very convenient way of representing complex numbers is by using their polar form. In this form, the distance r from the point associated with the complex number in the complex plane to the origin (the point $(0,0)$), and the angle φ between the vector associated with the complex number and the positive real axis are used. The distance r can be calculated as follows (please refer to Fig. 1.4):

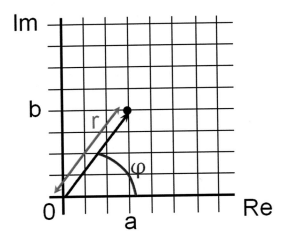

Fig. 1.4 Illustration of the polar form of the complex number a + bi in the complex plane. *Re* real axis, *Im* imaginary axis, *r* absolute value or modulus, φ argument.

$$r = \sqrt{a^2 + b^2} = \sqrt{z\bar{z}} \equiv |z|$$

Here, the symbol '$\equiv$' stands for 'is defined as' and we use the complex conjugate of z again. The symbol '|.|' stands for modulus or absolute value. The angle, or argument φ can be calculated by employing the trigonometric tangent function (see Chap. 3).

The polar expression of the complex number z is then (according to Euler's formula, see Sect. 3.3.1) given by:

$$z = re^{i\phi}$$

At this point, this may seem like a curious, abstract form of an exponential power and may seem not very useful. However, this polar form of complex numbers does allow to e.g., find all 3 complex roots of the equation $z^3 = 1$ and not just the one obvious real root $z = 1$ (see also Chap. 2 on equation solving and Sect. 3.3.1).

1.3   Mathematical Symbols and Formulas

The easiest way to learn the language of mathematics is to practice it, just like for any foreign language. For that reason we explain most symbols in this book in the context of how they are used. However, since mathematics is a very extensive field and since practicing mathematics takes time, we here also provide a more general introduction to and reminder of often used mathematical symbols and some conventions related to using the symbolic language of mathematics.

1.3.1 Conventions for Writing Mathematics

There are a few conventions when writing mathematical texts, that are also helpful to know when reading such texts. In principle, all mathematical symbols are written in Italics when they are part of the main text to discern them from non-mathematical text. Second, vectors and matrices (see Chaps. 4 and 5) are indicated in bold, except when writing them by hand. Since bold font can then not be used, (half) arrows or overbars are used above the symbol used for the vector or matrix. Some common mathematical symbols are provided in Table 1.3.

1.3.2 Latin and Greek Letters in Mathematics

To symbolize numbers that have no specific value (yet), both Latin and Greek letters are typically used in mathematics. In principle, any letter can be used for any purpose, but for quicker understanding there are some conventions on when to use which letters. Some of these conventions are provided in Table 1.4.

1.3.3 Reading Mathematical Formulas

To the less experienced, reading mathematical formulas can be daunting. Although practice also makes perfect here, it is possible to give some general advice on how to approach a mathematical formula and I will do so by means of an example. Suppose you are reading an article (Ünlü et al. 2006) and you stumble upon this rather impressive looking formula (slightly adapted for better understanding):

$$C_i^m(\varepsilon) = \frac{|\{(j,k)|(|r(i+k-1) - r(j+k-1)| \leq \varepsilon) \ for \ k=1\ldots m, j=i\ldots N-m+1\}|}{N-m+1}$$

The first thing to do when encountering a formula, is to make sure that you know what each of the symbols means in the context of the formula. In this case, I read the text to find out what C is (I already know that it will depend on m, i and ε from the left hand side of the

Table 1.3 Meaning of some common mathematical symbols with examples

Symbol	Meaning	Example
$\Rightarrow$	implies	$z = i \Rightarrow z^2 = -1$
$\Leftrightarrow$	if and only if	$x + 3 = 2x - 2 \Leftrightarrow x = 5$
$\approx$	approximately equal to	$\pi \approx 3.14$
$\propto$	proportional to	$y = 3x \Rightarrow y \propto x$
$!$	factorial	$3! = 3 \times 2 \times 1 = 6$
$<$	less than	$3 < 4$
$>$	greater than	$4 > 3$
$\ll$	much less than	$1 \ll 100,000,000$
$\gg$	much greater than	$100,000,000 \gg 1$

Table 1.4 Conventions on the use of Latin and Greek letters in mathematics

Latin letter	Application	Example
$a, b, c, \ldots$	as *parameter* in equations, or functions	$y = ax + b$ $y = ax^2 + bx + c$ $z = a + bi$
	vectors	$\mathbf{a}$ or $\vec{a}$ or $\overrightarrow{a}$
e	base of natural logarithm, approximately equal to 2.71828…	
x, y, z	*Cartesian coordinates*	$(x,y) = (1,3)$ $(x,y,z) = (-1,2,-4)$
	axes in 2D- or 3D space	x-axis
d, D	diameter	
	derivative (see Chap. 6)	$\frac{d}{dt}, \frac{d^2}{dx^2}$
i, j, k	counters	$i = 1, \ldots, n$ $\sum_{i=1}^{n} x_i$ $\sum_{i=1}^{n} \sum_{j=1}^{m} x_{i,j}$
	vector element	x_i
	matrix element	$x_{i,j}$
	complex unity	$z = a + bi$
n, m, N	quantity	$i = 1, \ldots, n$ $j = 1, \ldots, m$
	number of participants/animals in experimental science	N
P, Q, R	point in space	$P = (1,2)$ $Q = (-1,1,3)$
r	radius	circle or sphere radius
	modulus in polar coordinates or polar form of complex numbers	$z = re^{i\phi}$
t	time (counter)	
T	time (window), period	

Greek letter	Application	Example
α (alpha)	angle	
	significance level (in statistics)	
β (beta)	*power* (in statistics)	
δ (delta)	Dirac delta	$\delta(x) = \begin{cases} \infty & if\ x = 0 \\ 0 & if\ x \neq 0 \end{cases}$
	Kronecker delta	$\delta_{ij} = \begin{cases} 0 & if\ i \neq j \\ 1 & if\ i = j \end{cases}$
Δ (delta)	small increment	Δt
ε (epsilon)	(very) small (positive) number	for every $\delta < \varepsilon$
φ (phi)	angle (in polar coordinates)	
	argument (in polar form of complex numbers)	$z = re^{i\phi}$
ζ (zeta), θ (theta), ξ (ksi), ψ (psi)	angles	
π	relation between circumference and radius r of a circle	circumference $= 2\pi r$

formula), what i, j, and k count (I already know they will most likely be counters because of the mathematical convention explained in Table 1.4), what m and N are and what r is. I already know that ε will be a small number (again, because of the mathematical convention explained in Table 1.4). Then what remains to be known are the symbols .| (here: for every pair (j,k) such that), |.| (here: *cardinality* (the number of pairs (j,k); outer symbols) and distance (inner symbols)) and {.} (the set of).

What the article tells me is that $r(i)$ is a collection of m consecutive data points (an m-tuple) taken from a total number of N data points, starting at the ith data point and that $C_i^m(\varepsilon)$ is the relative number of pairs of these m-tuples which are not so different, i.e. which differ less than a small number ε for each pair of entries. $N - m + 1$ is the total number of different m — tuples that can be taken from the total data set. The largest value that i can take on is thus also $N - m + 1$. The first thing to notice in the formula is that by letting k run from 1 to m and j from i to $N - m + 1$, all possible pairs of m-tuples are indeed considered. This can be understood by assuming a value for m (e.g., 2), taking into account that i runs from 1 to $N - m + 1$ and then writing out the indices of the first and last pairs of m-tuples with indices $i + k - 1$ and $j + k - 1$. The part between the round brackets makes sure that from all possible pairs, only the pairs of m-tuples that have a distance smaller than ε are counted. So this formula indeed calculates what it is supposed to do.

So, what general lessons about formula reading can be learned from this example?

First, you need to know what all symbols mean in the context of the formula. Second, you need to understand what the more general mathematical symbols mean in the formula. Third, you break the formula into pieces and build up your understanding from there. More examples will be given in each of the following chapters.

Glossary

Arithmetic Operations between numbers, such as addition, subtraction, multiplication and division

Axis Reference line for a coordinate system

Base The number b in the exponentiation operation b^n

Cardinality The number of elements in a set, also indicated by #

Cartesian coordinates Uniquely specify a point in 2D space as a pair of numbers that indicates the signed distances to the point from two fixed perpendicular directed lines (axes), measured in the same unit of Length

Common denominator The least common multiple, i.e. the smallest number that can be divided by both denominators

Complex conjugate For a complex number $a + bi$ the complex conjugate is $a - bi$

Complex numbers Pairs of numbers (a,b) where a is the real part and b the imaginary part, also indicated as $a + bi$, where $i = \sqrt{-1}$ is the imaginary unit

Cryptography Discipline at the intersection of mathematics and computing science that deals with various aspects of information security

Denominator Lower part in a fraction $\frac{a}{b}$

Derivative Measure of change in a function

Distributive law To multiply a sum (or difference) by a factor, each element is multiplied by this factor and the resulting products are added (or subtracted)

Dyscalculia Difficulty learning or comprehending numbers or arithmetic, often considered as a developmental disorder

Exponent The number n in the exponentiation operation b^n

Finite decimal number Decimal number with a finite number of decimals

Imaginary Here: the imaginary part of a complex number

Infinite decimal number Decimal number with an infinite numbers of decimals

Infinity The number larger than any countable number

Integer numbers The numbers $\ldots, -3, -2, -1, 0, 1, 2, 3, \ldots$ collectively referred to as $\mathbb{Z}$

Irrational numbers Real numbers such as π that are not rational

Logarithm The base n logarithm of a number y is the exponent to which n must be raised to produce y

Natural logarithm Logarithm with base e

Natural numbers The numbers $0, 1, 2, 3, \ldots$ collectively referred to as $\mathbb{N}$

Numerator Upper part in a fraction $\frac{a}{b}$

Parameter Consider a function $f(x) = ax^2 + bx + c$; here, x is a variable and a, b, and c are parameters, indicating that the function represents a whole class of functions for different values of its parameters

Power Power is used in the context of exponentiation, e.g. as in b^n, where b is the base and n the exponent. One can also describe this as 'b is raised to the power n' or 'the nth power of b'

Prime numbers Natural numbers >1 that only have 1 and themselves as positive divisors

Rational numbers Numbers that can be written as fractions; a ratio of two integers, collectively referred to as $\mathbb{Q}$

Real numbers Numbers that can be measured but that cannot (always) be expressed as fractions, collectively referred to as $\mathbb{R}$

Root The nth root of a number x is the number y such that $y^n = x$.

Scientific notation Used to write numbers that are too large to be written in decimal form. In this notation all numbers are written as $a \times 10^b$, where a can be any real number and b is an integer

Significance level In statistics: probability of rejecting the null hypothesis given that it is true. Typically, the significance level is set to 0.05, meaning that a 1 in 20 chance of falsely rejecting the null hypothesis is accepted

Significant digit A digit with meaning for the number

Synesthete A person who, when one sense is stimulated, has experiences in another. In the context of this chapter: a person who identifies colors and shapes with numbers.

Vector Entity with a magnitude and direction, often indicated by an arrow, see Chap. 4

Symbols Used in This Chapter (in Order of Their Appearance)

The symbols presented in Tables 1.3 and 1.4 are not repeated here.

0,1,2,3,4,5,6,7,8,9	Arabic numerals used in the base 10 numeral system
$\ldots$	the pattern continues
$+$	addition
$=$	equal to
$\times$	multiplication
∞	infinity
$\circ$	degree
$\subset$	is a subset of
$\mathbb{N}$	natural numbers

$\mathbb{Z}$	integer numbers
$\mathbb{Q}$	rational numbers
$\mathbb{R}$	real numbers
$\mathbb{C}$	complex numbers
$\div$	division
$\sqrt{}$	square root
e	Euler's number
I	Roman numeral for 1
V	Roman numeral for 5
X	Roman numeral for 10
L	Roman numeral for 50
C	Roman numeral for 100
D	Roman numeral for 500
E	scientific notation for a base 10 exponent on calculators, e.g. 1E3 = 10^3
i	complex unity (positive solution of $x^2 + 1 = 0$)
$\bar{}$	complex conjugate (overbar)
r	here: modulus of complex number
φ	here: argument of complex number
$|\cdot|$	absolute value, modulus, cardinality or distance
$\equiv$	defined as
$\{.\}$	the set of
$.|$	such that
#	cardinality

Overview of Equations, Rules and Theorems for Easy Reference

<u>Relationship between numeral systems</u>
$$\mathbb{N} \subset \mathbb{Z} \subset \mathbb{Q} \subset \mathbb{R} \subset \mathbb{C}$$

<u>Order of arithmetic operations</u>

1) brackets (or parentheses)
2) exponents and roots
3) multiplication and division
4) addition and subtraction

<u>Arithmetic with fractions</u>

$$\frac{a}{b} \times \frac{c}{d} = \frac{ac}{bd}$$

$$\frac{a}{b} \div \frac{c}{d} = \frac{a}{b} \times \frac{d}{c} = \frac{ad}{bc}$$

Arithmetic with exponentials

$$a^0 = 1$$
$$a^{-n} = \frac{1}{a^n}$$
$$a^n a^m = a^{n+m}$$
$$\frac{a^n}{a^m} = a^{n-m}$$
$$(a^n)^m = a^{nm}$$
$$(ab)^n = a^n b^n$$

Arithmetic with logarithms

$$\log_b y^a = a\log_b y$$
$$\log_b \sqrt[a]{y} = \frac{\log_b y}{a}$$
$$\log_b xy = \log_b x + \log_b y$$
$$\log_b \frac{x}{y} = \log_b x - \log_b y$$
$$\log_b y = \frac{\log_k y}{\log_k b}$$

Arithmetic with complex numbers

$$(a + bi) + (c + di) = (a + c) + (b + d)i$$
$$(a + bi) - (c + di) = (a - c) + (b - d)i$$
$$(a + bi)(c + di) = (ac - bd) + (ad + bc)i$$
$$\frac{a + bi}{c + di} = \frac{ac + bd}{c^2 + d^2} + \frac{bc - ad}{c^2 + d^2} i \qquad (1.3)$$
$$\overline{a + bi} = a - bi$$

$z\bar{z} = a^2 + b^2$ for a complex number $z = a + bi$

Answers to Exercises

1.1. a) $\mathbb{Z}$ integer numbers
   b) $\mathbb{R}$ real numbers
   c) $\mathbb{R}$ real numbers
   d) $\mathbb{Q}$ rational numbers

e) $\mathbb{N}$ natural numbers

f) $\mathbb{R}$ real numbers

g) $\mathbb{C}$ complex numbers

1.2. a) $\frac{24}{21} = \frac{8}{7} = 1\frac{1}{7}$

b) $\frac{60}{48} = 1\frac{12}{48} = 1\frac{1}{4}$

c) $\frac{20}{7} = 2\frac{6}{7}$

d) $\frac{20}{6} = 3\frac{2}{6} = 3\frac{1}{3}$

1.3. a) $\frac{1}{3} + \frac{2}{5} = \frac{5}{15} + \frac{6}{15} = \frac{11}{15}$

b) $\frac{3}{14} + \frac{7}{28} = \frac{6}{28} + \frac{7}{28} = \frac{13}{28}$

c) $\frac{1}{2} + \frac{1}{3} + \frac{1}{6} = \frac{3}{6} + \frac{2}{6} + \frac{1}{6} = \frac{6}{6} = 1$

d) $\frac{3}{4} + \frac{7}{8} + \frac{9}{20} = \frac{30}{40} + \frac{35}{40} + \frac{18}{40} = \frac{83}{40} = 2\frac{3}{40}$

e) $\frac{1}{4} - \frac{5}{6} + \frac{3}{8} = \frac{6}{24} - \frac{20}{24} + \frac{9}{24} = -\frac{5}{24}$

f) $-\frac{1}{3} + \frac{1}{6} - \frac{1}{7} = -\frac{14}{42} + \frac{7}{42} - \frac{6}{42} = -\frac{13}{42}$

1.4. a) $\frac{2}{3} \times \frac{6}{7} = \frac{12}{21} = \frac{4}{7}$

b) $1\frac{2}{5} \times 1\frac{3}{7} = \frac{7}{5} \times \frac{10}{7} = \frac{70}{35} = 2$

c) $\frac{5}{6} \div \frac{6}{5} = \frac{5}{6} \times \frac{5}{6} = \frac{25}{36}$

d) $\frac{11}{13} \times \frac{2}{3} \div \frac{6}{13} = \frac{11}{13} \times \frac{2}{3} \times \frac{13}{6} = \frac{22}{18} = 1\frac{4}{18} = 1\frac{2}{9}$

e) $\frac{2}{4} \div 2 \times \frac{12}{48} = \frac{2}{4} \times \frac{1}{2} \times \frac{12}{48} = \frac{3}{48} = \frac{1}{16}$

1.5. a) $8 \div 4 - 1 \times 3^2 + 3 \times 4 = 2 - 9 + 12 = 5$

b) $(8 \div 4 - 1) \times 3^2 + 3 \times 4 = 9 + 12 = 21$

c) $(8 \div 4 - 1) \times (3^2 + 3) \times 4 = 1 \times 12 \times 4 = 48$

d) $(8 \div 4 - 1) \times (3^2 + 3 \times 4) = 1 \times (9 + 12) = 21$

1.6. a) $\frac{2^3 2^4}{2^2} = 2^{3+4-2} = 2^5$

b) $\frac{(2^2)^{\frac{1}{2}} 2^3}{2^{-4} 2^2} = 2^{1+3+4-2} = 2^6$

1.7. a) $\frac{\sqrt[3]{1000}}{\sqrt[4]{16}} = \frac{10}{2} = 5$

b) $\sqrt[4]{25}\sqrt{5} = 5^{2\cdot\frac{1}{4}}5^{\frac{1}{2}} = 5^{\frac{1}{2}+\frac{1}{2}} = 5$

c) $\sqrt{3y^8} = y^4\sqrt{3}$

d) $\frac{\sqrt[4]{9}}{\sqrt[4]{3}} = \frac{\sqrt[8]{81}}{\sqrt[8]{3}} = \sqrt[8]{\frac{81}{3}} = \sqrt[8]{27}$

e) $\sqrt[3]{x^{15}} = x^{\frac{1}{3}\cdot15} = x^5$

f) $\sqrt[7]{p^{49}} = p^{\frac{1}{7}\cdot49} = p^7$

g) $\sqrt[3]{\frac{a^6}{b^{27}}} = \frac{a^{\frac{1}{3}\cdot6}}{b^{\frac{1}{3}\cdot27}} = \frac{a^2}{b^9}$

h) $\sqrt[3]{\frac{-27x^6y^9}{64}} = \frac{-3x^2y^3}{4}$

1.8. a) $\frac{\log_b\left([x^2+1]^4\right)}{\log_b\sqrt{x}} = \log_b\left([x^2+1]^4\right) - \log_b\sqrt{x} = 4\log_b(x^2+1) - \frac{1}{2}\log_b x$

b) $\log_2(8\cdot2^x) = \log_2 8 + \log_2 2^x = 3 + x$

c) $\frac{1}{\log_{27}3} = \frac{\log_{27}27}{\log_{27}3} = \log_3 27 = 3$

d) $\log_2\left(8\cdot\sqrt[3]{8}\right) = \log_2 8 + \log_2 8^{\frac{1}{3}} = 3 + \frac{1}{3}\log_2 8 = 3 + \frac{1}{3}\cdot3 = 4$ or $\log_2\left(8\cdot\sqrt[3]{8}\right) = \log_2(8\cdot2)$
   $= \log_2 16 = 4$

1.9. a) $\log_2 x^2 + \log_2 5 + \log_2\frac{1}{3} = \log_2\left(x^2\cdot5\cdot\frac{1}{3}\right) = \log_2\left(\frac{5}{3}x^2\right)$

b) $\log_3\sqrt{a} + \log_3(10) - \log_3 a^2 = \log_3\left(\frac{\sqrt{a}\cdot10}{a^2}\right) = \log_3\frac{10}{a\sqrt{a}}$

c) $\log_a a^2 - \log_a 3 + \log_a\frac{1}{3} = \log_a\frac{a^2}{3\cdot3} = \log_a\frac{a^2}{9}$

d) $\log_x\sqrt{x} + \log_x x^2 + \log_x\frac{1}{\sqrt{x}} = \log_x\frac{x^2\sqrt{x}}{\sqrt{x}} = \log_x x^2 = 2$

1.10. a) $10 = 2 + 0 = 2$
   b) $111 = 4 + 2 + 1 = 7$
   c) $1011 = 8 + 0 + 2 + 1 = 11$
   d) $10101 = 16 + 0 + 4 + 0 + 1 = 21$
   e) $111111 = 32 + 16 + 8 + 4 + 2 + 1 = 63$
   f) $1001001 = 64 + 0 + 0 + 8 + 0 + 0 + 1 = 73$

1.11. a) 5.4×10^4
   b) 3.6×10^{-3}
   c) 1×10^2
   d) 1×10^{-5}
   e) 6.543×10^5
   f) 7.42×10^{-10}

1.12. The numbers a) to d) have been drawn in this figure

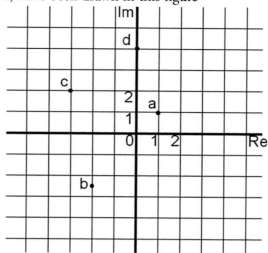

1.13. a) $(1 + i) + (-2 + 3i) = -1 + 4i$
   b) $(1.1 - 3.7i) + (-0.6 + 2.2i) = 0.5 - 1.5i$
   c) $(2 + 3i) - (2 - 5i) = 8i$
   d) $(4 - 6i) - (6 + 4i) = -2 - 10i$
   e) $(2 + 2i) \times (3 - 3i) = 6 + 6i - 6i - 6i2 = 12$
   f) $(5 - 4i) \times (1 - i) = 5 - 4i - 5i + 4i2 = 1 - 9i$
   g) $\dfrac{5 - 10i}{1 - 2i} = \dfrac{(5 - 10i)(1 + 2i)}{1^2 + 2^2} = \dfrac{5 - 10i + 10i - 20i^2}{5} = 5$
   h) $\dfrac{18 + 9i}{\sqrt{5} - 2i} = \dfrac{(18 + 9i)(\sqrt{5} + 2i)}{5 + 2^2} = \dfrac{18\sqrt{5} + 9\sqrt{5}i + 18i + 18i^2}{9}$
   $$= (2\sqrt{5} - 2) + (\sqrt{5} + 2)i$$

References

Online Sources of Information: Methods

https://en.wikipedia.org/wiki/Numbers
https://en.wikipedia.org/wiki/Arithmetic
https://en.wikipedia.org/wiki/List_of_mathematical_symbols
www.khanacademy.org/math

Online Sources of Information: Others

http://sliderulemuseum.com/

Books

A.S. Posamenter, B. Thaller, *Numbers: Their Tales, Types and Treasures* (Prometheus Books, Amherst, NY, 2015)

Papers

A. Ünlü, R. Brause, K. Krakow, Handwriting analysis for diagnosis and prognosis of Parkinson's disease, in *ISBMDA 2006, LNBI 4345*, ed. by N. Maglaveras et al. (Springer, Berlin, 2006), pp. 441–450

2

Equation Solving

Branislava Ćurčić-Blake

After reading this chapter you know:

- what equations are and the different types of equations,
- how to solve linear, quadratic and rational equations,
- how to solve a system of linear equations,
- what logarithmic and exponential equations are and how they can be solved,
- what inequations are and
- how to visualize equations and solve them graphically.

2.1 What Are Equations and How Are They Applied?

An *equation* is a mathematical expression; a statement that two quantities are equal. A simple example is given by

$$5 = 5$$

or

$$3 + 2 = 5$$

These are equations without *unknowns*. For statements like these to be true the values of the expressions on each side of the equal sign have to be the same. Often, equations have one *variable* that is unknown, such as

---

B. Ćurčić-Blake (✉)
Neuroimaging Center, University Medical Center Groningen, Groningen, The Netherlands
e-mail: b.curcic@umcg.nl

© Springer International Publishing AG 2017
N. Maurits, B. Ćurčić-Blake, *Math for Scientists*, DOI 10.1007/978-3-319-57354-0_2

$$3 + x = 5$$

Here, variable x is unknown, and to solve the equation, we have to find its value so that the above equation becomes true. The solution is $x = 2$, because if we substitute it into the above equation, it becomes

$$3 + 2 = 5$$

which is true. Solving equations dates back several thousands of years. For example, the Babylonians (2000–1000 BC) already used equation solving to calculate the dimensions of a rectangle given its surface and the difference between its height and width.

2.1.1 Equation Solving in Daily Life

You may not be aware that we use equations every day. Often, equation solving is involved when dealing with money, e.g. when one needs to calculate percentages, differences or taxes. An example is the following:

Example 2.1

Marc wants to buy three beers (€1.50 each) and one bottle of wine (€10.00). He has €15.00. How much money does he have left after he finished shopping?

If we denote the change with x, we can write the given information in the form of an equation as follows:

$$3 \cdot 1.5 + 1 \cdot 10 + x = 15$$

Here, on the left side all expenses and change x are included. Together, they need to add up to the €15.00 that Marc has. The equation can be solved in two steps by first adding all like terms:

$$14.5 + x = 15 \quad \rightarrow \quad x = 0.5$$

Thus if Marc buys all the drinks he wants, he will have €0.50 left.

In everyday shopping we thus use equations, without even thinking about it. A similar example is provided by calculating sale prices:

Example 2.2

The bag that originally cost €70.00 is now on sale at a 25% discount. What is the sale price of the bag?

If the new price is denoted as x we can compose and solve an equation as follows:

$$x = 70 - 70 \cdot \frac{25}{100} \rightarrow x = 70(1 - 0.25) \rightarrow x = 70 \cdot 0.75 \rightarrow x = 52.5$$

Thus, the sale price of the bag is €52.50.

In this section some specific examples of equations were provided, to get you introduced to the topic. In the next section, we will generalize these examples and introduce some definitions associated with equations.

2.2 General Definitions for Equations

2.2.1 General Form of an Equation

Linear equations in one unknown, such as the examples above, can generally be written as:

$$ax = b, \tag{2.1}$$

where x is the unknown or the variable that we aim to solve the equation for and a and b are constants.

2.2.2 Types of Equations

The equations that were introduced as examples in Sect. 2.1 were quite simple and were examples of linear equations with one unknown (x). More generally, we can distinguish between *linear* and *nonlinear* equations that have one or more unknowns.

A linear equation with one unknown is an equation that can be rewritten to the form of Eq. (2.1). It only includes terms that are constant or that are the product of a constant and a single variable to its first power. Interestingly, such an equation describes a straight line. You can think of linear as straight in two dimensions or flat (as a plane) in three dimensions. Linear equations in two unknowns can be represented by:

$$ax + by + c = 0,$$

where x and y are the unknowns or the variables that we aim to solve the equation for and a, b and c are constants. To find a solution for both variables, a minimum of two *independent* equations is necessary.

Nonlinear (*polynomial*) equations are equations in which one or more terms contain a variable with a power different from one and/or in case of a nonlinear equation with more unknowns there is a term with a combination of different variables. Such a combination could e.g. be a product or a quotient of variables. Other non-polynomial nonlinear equations will be introduced in Sects. 2.6 and 2.7.

2.3 Solving Linear Equations

In Sect. 2.1.1 we already solved some equations without thinking too much about our approach. This can often be done, as long as the equations are very simple. However, if equations become a bit more complicated, it is useful to have a recipe for equation solving that always works, as long as a solution exists, of course. The general goal of solving any linear equation with one unknown x is to get it into the form

$$x = c, \tag{2.2}$$

where c is a constant. ANY linear equation can be solved, i.e. transformed into the form of Eq. (2.2) by following the general rules listed below:

1. Expand terms, when the equation is not (yet) written as a sum of linear terms.
2. Combine like terms by adding or subtracting the same values from both sides.
3. Clear out any *fractions* by multiplying every term by the *least common denominator*.
4. Divide every term by the same <u>non-zero</u> value to make the constant in front of the variable in the equation equal to 1.

These rules may sound rather abstract now, but in the next sections, some of these rules are explained in more detail and you can practice their application in some exercises.

2.3.1 Combining Like Terms

To solve a linear equation in one unknown one of the first steps is to combine like terms, which means that all terms with the same variable (e.g. x) are gathered. In other words, if you start with an equation

$$ax + b = c + dx$$

you will gather all terms in x on the same side to rewrite it to:

$$(a - d)x = c - b$$

Here, to transfer a term to the other side of the equality sign, we had to subtract it from the side we transfer it from. To keep the equation true, we thus also have to subtract that term from the side of the equation we transfer it to. This means that when you are transferring a term from one side of the equation to the other you are basically changing its sign. In this case, by subtracting dx from the right hand side of the equation, we also had to subtract it from the left hand side of the equation, where it thus got a negative sign. Something similar happened to the constant term b.

Example 2.3

Sjoerd ordered three bottles of wine online. Postage was €9.00 and the total costs were €45.00. How much did each bottle of wine cost?

    To solve this problem we represent the price of a bottle of wine by x. Then we can compose an equation as follows:

(continued)

Example 2.3 (continued)

$$3x + 9 = 45$$

Here the like terms are 9 and 45 because they are both constants. So we combine them:

$$3x = 45 - 9 \quad \rightarrow \quad 3x = 36 \quad \rightarrow \quad x = 12$$

We find that each bottle of wine costs €12.00.

Exercise

2.1. Grandma Jo left 60 pieces of rare coins to her heirs. She had two daughters. However, the older of the two daughters, Mary, died just a few days before Grandma Jo, so her coins are to be divided among Mary's three daughters. How many rare coins will each granddaughter of Grandma Jo inherit?

2.2. Solve the following linear equations in one unknown:

a) $7x + 2 = -54$
b) $-5x - 7 = 108$
c) $3x - 9 = 33$
d) $5x + 7 = 72$
e) $4x - 6 = 6x$
f) $8x - 1 = 23 - 4x$

2.3.2 Simple Mathematical Operations with Equations

Solving an equation may be fun, if you look at it as solving a puzzle. You may swap 'pieces' around, and you may try to fit 'pieces' in different ways. Often you may have to think out of the box and find creative solutions. The easiest way is to perform simple mathematical operations such as addition, subtraction, multiplication or division, as described in rules 2 and 3 above. In case of solving a linear equation with one unknown you may want to e.g. add a constant or multiply the equation by a constant. Consider the following example:

$$\frac{5}{3}x + \frac{1}{3} = 5$$

In this case it might help to first multiply the whole equation by 3 to get rid of the fractions. To keep the equation true, you have to multiply both sides by the same number:

$$\frac{5}{3}x + \frac{1}{3} = 5 \quad \Big|\cdot 3 \quad \rightarrow \quad 5x + 1 = 5 \cdot 3 \rightarrow \quad 5x = 15 - 1 \quad \rightarrow \quad x = \frac{14}{5}$$

Note that we also applied rule number 4. Similarly, to solve

$$100x + 50 = 200$$

you may first want to divide by 100:

$$100x + 50 = 200 \quad \left| \cdot \frac{1}{100} \right. \quad \rightarrow \quad x + \frac{1}{2} = 2 \quad \rightarrow \quad x = \frac{3}{2}$$

In Box 2.1 a useful set of rules to solve equations is given, summarizing more formally what we just did.

Box 2.1 Useful *Arithmetic* Rules for Solving Linear Equations

If $a = b$ and $c \in \mathbb{C}$ then

$$a + c = b + c$$
$$ac = bc$$

If $a = b$, $c \in \mathbb{C}$ and $c \neq 0$ then

$$\frac{a}{c} = \frac{b}{c}$$

If F is any *function* and $a = b$ then

$$F(a) = F(b)$$

Exercise

2.3. Solve the following equations:
   a) $\frac{2x}{3} = \frac{x}{3}$
   b) $\frac{2x}{3} = 5x + 3$

2.4   Solving Systems of Linear Equations

We briefly mentioned, when we introduced linear equations with two unknowns, that to find a solution for both variables, a minimum of two independent equations is necessary. Such a set of linear equations is the smallest possible system of linear equations. In general, a system of linear equations is a set which contains two or more linear equations with two or more unknowns. This might sound rather abstract, but actually such a system is regularly encountered in practice.

Example 2.4

Suppose John only ate bread with peanut butter or jam during 1 month. Each jar of jam contains 30 g of sugar and 20 g of other nutrients. Each jar of peanut butter contains 15 g of sugar and 40 g of other nutrients. How many jars of jam and peanut butter did John use in this month if he consumed 150 g of sugar and 450 g of other nutrients?

(continued)

Example 2.4 (continued)

We can rewrite the problem as a system of two linear equations in two unknowns. Suppose that x is the number of jars of jam, and y is the number of jars of peanut butter, then the system of linear equations describing this problem is:

$$150 = 30x + 15y$$

$$450 = 20x + 40y$$

This system of equations has the solution $x = -\frac{5}{6}$, $y = 11\frac{2}{3}$, which you can for now verify by substituting it in the system of equations.

Here is another example, in which a system of three linear equations in three unknowns can help solve a practical problem encountered when three friends have dinner together.

Example 2.5

Mary, Jo and Sandy had dinner together and agree to go Dutch and split the bill according to what they had. They received two bills, one for drinks (total of €26.00) and one for food (€36.00) without further details. Mary had two glasses of wine, one juice and salmon. Jo had three glasses of wine and salmon. Sandy had beefsteak and two glasses of juice. They remember that juice cost €2.00 a glass, and that beefsteak was €3.00 more expensive than salmon. How much does each colleague have to pay?

We thus need to determine what wine (x), salmon (y) and beefsteak (z) cost. We know that beefsteak was €3.00 more expensive than salmon, which will give us the first equation. Further, we know that the three of them together ate two salmons and one beefsteak, giving us the second equation, and had five glasses of wine and three glasses of juice, giving us the third equation. We can thus write:

$$z = y + 3$$

$$2y + z = 36$$

$$5x + 3 \cdot 2 = 26$$

In total, these three equations are a system of three linear equations with three unknowns. The third equation is straightforward to solve as it has only one unknown x:

$$5x = 26 - 6 \quad \rightarrow \quad 5x = 20 \quad \rightarrow x = 4$$

Further, we can substitute the first equation $z=y+3$ into the second:

$$2y + (y + 3) = 36 \quad \rightarrow \quad 3y + 3 = 36 \quad \rightarrow \quad 3y = 33 \quad \rightarrow \quad y = 11$$

And if we now substitute y back into the first equation we get the price of the beefsteak

$$z = 11 + 3 = 14$$

(continued)

Example 2.5 (continued)

Now, we can calculate how much Mary, Jo and Sandy each have to pay:

$$Mary : 2x + 2 + y = 2 \cdot 4 + 2 + 11 = 21$$
$$Jo : 3y + x = 3 \cdot 4 + 11 = 23$$
$$Sandy : z + 2 \cdot 2 = 14 + 4 = 18$$

As a final check we should verify whether the total bills sum up to the same total amount. The two bills for drinks and food add up to €36.00 + €26.00 = €62.00. The bills for Mary, Jo and Sandy add up to €21.00 + €23.00 + €18.00 = €62.00. Thus, the final bills match and we solved the problem.

The approach that we followed for solving the problem in Example 2.5 is called *substitution*, which is a bit of an ad-hoc method. There are also more systematic *algebraic* methods of solving systems of linear equations and a graphical one. The methods for solving systems of linear equations that we explain in the following sections in more detail are the following:

a. Solving by substitution
b. Solving by elimination
c. Solving graphically
d. Solving by Cramer's rule

2.4.1 Solving by Substitution

This is the basic method to solve a system of any type of equations. It is not always easy to implement, but if possible gives straightforward answers. Basically, if you have a system of n equations with n unknowns, you first solve one equation to obtain the solution for one variable, then substitute that solution in the next equation to obtain the solution for the next variable etcetera. This procedure is further illustrated in the next examples for systems of linear equations.

Example 2.6

Solve the following system of two linear equations in two unknowns:

$$3x + 5 = 5y$$
$$2x - 5y = 6$$

To solve this system, you can start by e.g. expressing y in terms of x based on the first equation: $y = \frac{3}{5}x + 1$ and then substitute y into the second equation: $2x - 5(\frac{3}{5}x + 1) = 6$, to solve this equation:

$$2x - 3x - 5 = 6 \quad \rightarrow -x = 11 \quad \rightarrow \quad x = -11$$

(continued)

Example 2.6 (continued)

The solution for x can then be substituted into the expression for y to find its solution:

$$y = \frac{3}{5}x + 1 = \frac{-33}{5} + 1 \quad \to y = \frac{-28}{5} = -5\frac{3}{5}$$

A similar procedure can be followed for a system of more than two unknowns:

Example 2.7

Solve the following system of three linear equations in three unknowns:

$$2x + y + z = 4$$
$$x - 7 - 2y = -3z$$
$$2y + 10 - 2z = 3x$$

We can solve this system by substitution e.g., by first obtaining x from the second equation:

$$x = 2y - 3z + 7$$

We can now substitute x into the first and third equations and gather like terms:
First equation: $2(2y-3z+7)+y+z=4 \to 4y-6z+14+y+z=4 \to 5y-5z=-10$
Third equation: $2y+10-2z=3(2y-3z+7) \to 2y+10-2z=6y-9z+21 \to -4y+7z=11$
Thus, now we have two equations in two unknowns (y and z):

$$5y - 5z = -10 \quad \to y = z - 2$$
$$-4y + 7z = 11$$

Finally, we substitute the expression for y in the last equation: $-4(z-2)+7z=11 \to 3z=3$. Thus $z=1$ and by substituting this in the other equations, we find that $y=-1$ and $x=2$. To verify whether this solution is correct, we can insert the values of x, y and z into the original system of equations to see if the equations become true.

$$2x + y + z = 4 \quad \to \quad 2 \cdot 2 - 1 + 1 = 4 \quad \to \quad 4 = 4 \quad True$$
$$x - 7 - 2y = -3z \quad \to \quad 2 - 7 - 2 \cdot (-1) = -3 \cdot 1 \quad \to \quad -5 + 2 = -3 \quad True$$
$$2y + 10 - 2z = 3x \quad \to \quad 2 \cdot (-1) + 10 - 2 \cdot 1 = 3 \cdot 2 \quad \to \quad 6 = 6 \quad True$$

We have thus found a correct solution for the system of equations.

Exercises

2.4. Solve the following systems of linear equations ($\wedge$ means 'and'):

a) $x - 2y = 4 \wedge \frac{x}{3} - y = \frac{4}{3}$

b) $2x - 2y = 5 \wedge -4x + 8y = -16$

c) $\frac{2x}{3} - y = 2 \wedge x + \frac{y}{2} = 1$

d) $0.2x - 0.4y = 0.6 \wedge -0.8x + 1.8y = -2.4$

2.4.2 Solving by Elimination

As illustrated in the previous section, solving a system of more than two linear equations by substitution can be very lengthy and clumsy. Mistakes are almost guaranteed. A more elegant way, which is sometimes—but not always—easier to implement, is to solve a system of linear equations by elimination. In essence, one then tries to manipulate one equation (e.g. by multiplication or division by a constant) to make the coefficients of one of the variables the same in two equations. Then you can subtract or add these two equations to eliminate that particular variable. Let's make this more clear by an example:

Example 2.8

Consider the following system of two linear equations in two unknowns:

$$x + y = 10$$
$$x - y = 2$$

As you can see the coefficient of x is the same in both equations. This implies that by simply subtracting the second equation from the first we can eliminate x. We do that by subtracting the terms on the left side from each other, and the terms on the right side, separately!

$$x - x + y - (-y) = 10 - 2 \rightarrow 0 + 2y = 8 \rightarrow y = 4$$

We can now substitute y in any of the two equations. If we substitute it in the first, we find that:

$$x + 4 = 10 \quad \rightarrow \quad x = 6$$

You can now verify yourself whether we found the correct solution by substituting x and y in the second equation, similar to how this was done in Example 2.7. Note that this system can also be solved by first eliminating y by adding the two equations.

In the previous example we did not have to manipulate any of the equations. However, often this is necessary to simplify the procedure, as illustrated in the next example:

Example 2.9

Consider the following system of two linear equations in two unknowns:

$$2x + y = 4$$
$$x - 3y = -1$$

To solve this system we first multiply the second equation by -2:

$$2x + y = 4$$
$$-2x + 6y = 2$$

Then we add the two equations to find that:

$$7y = 6$$

Thus, $y = \frac{6}{7}$ and $x = -1 + 3 \cdot \frac{6}{7} = -\frac{7}{7} + \frac{18}{7} = \frac{11}{7}$. Of course, this system can also be solved by choosing other elimination strategies.

When solving a system of three linear equations in three unknowns, in the first step one should eliminate one variable by manipulating two of the three equations as illustrated in the next example:

Example 2.10

Consider the following system of three linear equations in three unknowns:

$$x + y + z = 3$$
$$2x + 4y - z = -7$$
$$-3x + 2y + z = -7$$

This system can be solved by first eliminating y, for example by first multiplying the first equation by 2 and dividing the second equation by 2. This gives the following system of equations:

$$2x + 2y + 2z = 6$$
$$x + 2y - \frac{1}{2}z = -3\frac{1}{2}$$
$$-3x + 2y + z = -7$$

In the next step, we reduce the system of three equations, to a system of two equation by eliminating y. We do this by subtracting the second equation from both the first and the third equation:

$$x + 2\frac{1}{2}z = 9\frac{1}{2}$$
$$-4x + 1\frac{1}{2}z = -3\frac{1}{2}$$

(continued)

Example 2.10 (continued)

Now we have a system of two equations with two variables, and we know how to solve that. We can do this e.g., by elimination again, by multiplying the first equation by 4 and then adding the two equations. This gives $z=3$. The values for x and y can then be found by substitution in other suitable equations to be $x=2$ and $y=-2$.

Exercise

2.5. Solve the following systems of linear equations:

 a) $7x + 4y = 2 \wedge 9x - 4y = 30$
 b) $2x + y = 4 \wedge x - y = -1$
 c) $2x - 5y = 5 \wedge -6x + 7y = -39$
 d) $x - 2y + 3z = 7 \wedge 2x + y + z = 4 \wedge -3x + 2y - 2z = -10$

2.4.3 Solving Graphically

Each linear equation in two unknowns describes a straight line in the two-dimensional plane. Notice that a straight line in the (x,y)-plane is described by $y = ax + b$, where a is a coefficient describing the steepness of the line and b indicates the intersection with the y-axis. Sometimes it can help to solve a system of two linear equations in two unknowns by plotting the associated lines. To solve the system, you just have to find the intersection of the two lines.

Example 2.11

Consider the following system of two linear equations in two unknowns that was introduced in Example 2.8:

$$x + y = 10$$
$$x - y = 2$$

The associated curves are:

$$y = -x + 10$$
$$y = x - 2$$

The graph looks like this:

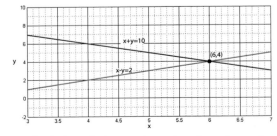

As the lines intersect at (6,4) the solution to the system is $x=6$, $y=4$.

Such a plot can also illustrate when a system of two linear equations in two unknowns has no solution: when the two associated lines run in parallel, there is no intersection. Of course, plotting the associated lines to find a solution to a system of equations only works when the plotting range is chosen such that the intersection is actually within the plotted part of the plane.

In principle, a graphical approach to find the solution of a system of three linear equations in three unknowns would also work, but this is not very practical. In this case, one would be looking for the intersection of three planes, which is much harder to plot and visualize.

2.4.4 Solving Using Cramer's Rule

One of the most practical ways of solving systems of linear equations is Cramer's rule. It offers a solution to a system of n equations in n unknowns by using *determinants* that are explained in detail in Sect. 5.3.1.

In its most simple form Cramer's rule can be applied to solve a system of two equations in two unknowns, according to the rule indicated in Box 2.2.

Box 2.2 Cramer's Rule for a System of 2 Linear Equations

The system of two equations with two unknowns x and y

$$ax + by = c$$
$$dx + ey = f$$

has the solution

$$x = \frac{ce - bf}{ae - bd}$$

$$y = \frac{af - cd}{ae - bd}$$

when $ae - bd \neq 0$.

Cramer's rule also applies to systems of n linear equations in n unknowns for $n > 2$, which is explained in Sect. 5.3.

2.5 Solving Quadratic Equations

Not all polynomial equations are linear. If a variable in a polynomial equation has power >1 the equation is nonlinear. Specifically, a quadratic equation in one unknown is a second-order polynomial equation of the form:

$$ax^2 + bx + c = 0,$$

where $a \neq 0$. Quadratic equations can be used, for example, to model relationships between variables when a linear relationship is not appropriate, e.g. when modeling walking speed as a function of age. Walking speed is slow when you are very young, reaches a peak and then decreases again when you get older. Its curve would thus show an inverted U-shape, i.e. a second-order polynomial shape. Statistical programs such as SPSS provide the option to do regression analysis (see also Sect. 4.3.2 for an explanation of linear regression) and model such nonlinear relationships.

It is important to note that solutions of quadratic equations are not always *unique*—meaning that there might be more than one solution. In fact, quadratic equations can have no, one or two solutions. For example, the equation $x^2 - 4 = 0$ has two solutions: $x = 2$ and $x = -2$. More generally, a quadratic equation can be solved by following the general rules outlined in Sect. 2.3, extended with the following rules that will be explained in more detail below:

1. If needed, apply a *function* to both sides (e.g. take the square root of both sides to get rid of a square).
2. Recognize a pattern you have seen before, like the difference of squares, or square of differences.
3. Factorize in order to simplify.
4. Equate each factor to zero.

Regarding rule 1, it is useful to remember that if F is any kind of function (such as the square or square root function, or the logarithmic function), and $a = b$, then also $F(a) = F(b)$. For example, if $a = b$, then $\log a = \log b$, $c^a = c^b$ and $e^a = e^b$ (see Box 2.1).

Example 2.12

Solve the equation:

$$2^{3x} = 4$$

This type of equation often occurs in chemistry calculations, for instance when calculating drug doses (see Sect. 2.9 for an example). To solve it we will take the logarithm with base 2 of the whole equation and apply arithmetic rules for logarithms (Sect. 1.2.2). This yields

$$\log_2 2^{3x} = \log_2 4 \quad \rightarrow \quad 3x = 2 \quad \rightarrow \quad x = \frac{2}{3}$$

Example 2.13

Solve the equation:

$$\sqrt{3x + 5} = 7$$

To solve this equation both sides can be squared. This yields:

$$3x + 5 = 7^2 \quad \rightarrow \quad 3x = 49 - 5 \quad \rightarrow \quad x = \frac{44}{3}$$

We will now consider three different ways of solving quadratic equations.

2.5.1 Solving Graphically

To get a better understanding of quadratic equations, we first discuss the graphical approach to solving them. This is done by first drawing the function $f(x) = ax^2 + bx + c$ in the (x,y)-plane. Its curve is a parabola. The solution of the equation is given by the points on the curve that intersect with the x-axis, since there $f(x) = 0$. When the parabola's peak just touches the x-axis, there is one solution to the equation, when the x-axis cuts the parabola into three parts, there are two solutions and in all other cases there are none.

The parabola associated with $f(x)$ has a maximum or a minimum, depending on the sign of a (illustrated in Fig. 2.1)

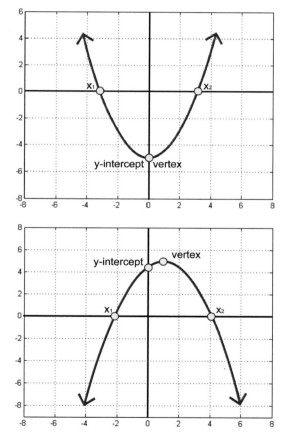

Fig. 2.1   Example of parabolas associated with quadratic equations. *Top*: The solutions indicated by x_1 and x_2 of the equation $0.5x^2 - 5 = 0$. Here $a > 0$, thus the curve is concave up. *Bottom*: The solutions indicated by x_1 and x_2 of the equation $-0.5(x-1)^2 + 5 = 0$. Here $a < 0$, thus the curve is concave down. For both parabolas, the *vertex* and the *y-intercept* are also indicated.

1. If $a > 0$, $f(x)$ is facing up (*concave* up) and the parabola has a minimum
2. If $a < 0$, $f(x)$ is facing down (concave down) and the parabola has a maximum
3. In any case, the peak of the parabola (or vertex) is at the position

$$x = \frac{-b}{2a}$$

2.5.2 Solving Using the Quadratic Equation Rule

Solving equations graphically, although intuitive, might be difficult, and not always exact, depending on whether the solution is a grid point. One method of solving quadratic equations that always works is the quadratic equation rule (Box 2.3):

Box 2.3 Quadratic Equation Rule

The solution of

$$ax^2 + bx + c = 0$$

is given by

$$x_{1,2} = \frac{-b \pm \sqrt{b^2 - 4ac}}{2a}$$

Here, the solution consists of two *roots*, x_1 and x_2, where the first is found for the plus sign and the second for the minus sign. Applying this rule is relatively easy, as illustrated in this example:

Example 2.14

Solve the following equation by applying the quadratic equation rule:

$$2x^2 + 3 = -5$$

Here $a = 2, b = 0$ and $c = 3 + 5 = 8$. By applying the quadratic equation rule we find the two solutions

$$x_1 = \frac{-b + \sqrt{b^2 - 4ac}}{2a} = \frac{0 + \sqrt{0 - 4 \cdot 2 \cdot 8}}{2 \cdot 2} = \frac{\sqrt{-64}}{4} = \frac{8i}{4} = 2i$$

$$x_2 = \frac{-b - \sqrt{b^2 - 4ac}}{2a} = \frac{0 - \sqrt{0 - 4 \cdot 2 \cdot 8}}{2 \cdot 2} = \frac{\sqrt{-64}}{4} = \frac{-8i}{4} = -2i$$

As you can see there are two solutions for this equation and both are complex numbers (see Sect. 1.2.4).

Exercise

2.6. Find the roots of the following equations using the quadratic equation rule:

a) $2x^2 + 7x + 5 = 0$
b) $x^2 - 9 = 0$
c) $x^2 - 3 = 5x + 2$
d) $x(x - 5) = 3$
e) $2(y^2 - 6y) = 3$
f) $1 + \frac{6}{x} + \frac{9}{4x^2} = 0$

2.5.3 Solving by Factoring

In my opinion this is the most interesting way of solving equations as it requires bringing out one's creativity and imagination. The goal is to recognise factors (simpler forms) and rewrite the equation using only a product of these simple forms, which are subsequently each set to zero. This is not always possible, but often it is. Factors are (in the case of an n-th order polynomial equation) simpler polynomials than the original equation. For example, quadratic equations should be factored into two first-order polynomials.

Example 2.15

Consider the quadratic equation $x(x-1)=0$ which is already factored into two factors: the factor x and the factor $(x - 1)$. To find the solution of this equation, at least one of the factors has to be set to zero. Hence, solutions are found by equating each of the factors to zero: $x=0$ OR $x-1=0$. Thus, the solutions are $x_1=0$ and $x_2=1$.

To solve polynomial equations by factoring, it is important to remember the following rule and special cases regarding multiplication of such factors:

Box 2.4 Factor Multiplication Rule

General rule:

$$(x + a)(x + b) = x^2 + (a + b)x + ab$$

Special cases:

$$(x + y)^2 = x^2 + 2xy + y^2$$
$$(x - y)^2 = x^2 - 2xy + y^2$$
$$x^2 - y^2 = (x - y)(x + y)$$

Example 2.16

Solve the following equation by factoring:

$$x^2 - 5x - 24 = 0$$

Thus, we need to find numbers a and b such that $(x-a)(x-b)=x^2-5x-24=0$. There are many different approaches to find these numbers. First, we will here use that if we add and subtract the same number we do not change the outcome:

$$x^2 - 5x - 24 = x^2 - 5x - 3x + 3x - 24$$

Now, we can rewrite the equation as

$$x^2 - 8x + 3x - 24 = 0,$$

allowing to take the factor $x - 8$ out which results in the other factor being $x + 3$:

$$x(x - 8) + 3(x - 8) = 0$$
$$(x + 3)(x - 8) = 0$$

Thus, the solutions are $x_1 = -3$ and $x_2 = 8$.

Example 2.17

Solve the following equation by factoring:

$$x^2 - 9 = 0$$

Here you have to realize that 9 is a square of 3 ($3^2 = 9$), and to remember one of the special cases of the factor multiplication rule (Box 2.4). We can then rewrite the left-hand side of the equation, which is a difference of squares, to:

$$x^2 - 9 = (x - 3)(x + 3)$$

Thus, the equation has two solutions $x-3=0$ or $x+3=0$; thus $x_1=3$ and $x_2=-3$.

Example 2.18

Solve the following equation by factoring:

$$x^2 + 6x + 9 = 0$$

If we realize that $3^2=9$ and that $3+3=6$, we see that the above equation is the square of a sum (a special case of the factor multiplication rule again). Thus we can rewrite the left-hand side of the equation to:

(continued)

Example 2.18 (continued)

$$x^2 + 6x + 9 = (x + 3)^2$$

and the above equation to:

$$(x + 3)^2 = 0$$

Thus, the solutions are $x_{1,2} = -3$. For this equation there is only one (double) solution.

To find the factors of a quadratic equation, there is also a more systematic approach. Suppose that you are trying to solve the equation $x^2 + cx + d = 0$. Taking the general factor multiplication rule into account, we know that we are looking for two numbers a and b such that their sum is c and their product is d. Hence, assuming that the solutions of the equation are integers, we can make a table of all possible pairs of factors of d that are likely solutions and calculate their sum. The pair that has the right sum and product then provides two solutions. Let's illustrate this approach for Example 2.16:

Example 2.16 (continued)

Solve the following equation by factoring:

$$x^2 - 5x - 24 = 0$$

We first construct a table with integer factors of -24 that are likely candidates for factors and add their sums:

Factor 1	Factor 2	Sum
-2	12	10
2	-12	-10
-3	8	5
3	-8	-5
-4	6	2
4	-6	-2

Now, we immediately see that the pair of factors 3 and -8 has the right sum. Thus, the equation can be factored into $(x+3)(x-8) = 0$ and its solutions are $x_1 = -3$ and $x_2 = 8$.

Exercise

2.7. Find the roots of the following equations using factoring:

   a) $x^2 - 7x + 2 = -8$
   b) $x^2 - 4x + 4 = x$
   c) $x^2 + 2x - 8 = 0$
   d) $2x^3 - 14x^2 + 20x = 0$

2.6 Rational Equations (Equations with Fractions)

We already touched upon fractions in Sect. 1.2.1. Rational equations use fractions. These are equations that have a rational expression on one or both sides in which the unknown variable is in one or more of the denominators. For example, in the following equation, the variable x is in the denominator on the left side:

$$\frac{3}{x+2} = 7 + \frac{x}{3}$$

Such equations are used for example in percentage calculations, or when calculating the speed at which a computer job will be done, as illustrated in the next example.

Example 2.19

You need to do a heavy calculation on a two core PC to obtain a fit to your data. Normally, core 1 would take 6 h to fit the data. However, core 2 is faster and would take 3 h to fit the data. How long will it take to calculate the data fit using both cores of the PC?

This problem that requires P calculations can be solved using rational equations. First, we realize that core 1 takes 6 h to do P calculations. Thus it can do $\frac{P}{6}$ calculations per hour. Second, we know that core 2 takes 3 h to do P calculations. Thus it can do $\frac{P}{3}$ calculations per hour. Thus if we say that T is the total time needed for both cores to calculate the data fit together, we find that we need to solve a rational equation:

$$\frac{P}{6} + \frac{P}{3} = \frac{P}{T}$$

Solving this equation we find that

$$\frac{9}{18} = \frac{1}{T} \quad \rightarrow \quad T = 2$$

Thus, using rational equations we find that it takes 2 h to calculate the fit using both cores of the PC. Note that this problem can also be solved in other ways that do not involve rational equations.

2.7   Transcendental Equations

These are equations where at least one side contains a so called *transcendental* (i.e. non-algebraic) function. This is a whole world of equations, and usually they can be written in the form $F(x)=0$, where $F(x)$ can be any function. Some examples of functions, discussed in this book are logarithmic, trigonometric or exponential functions. Importantly, these equations are not always solvable, or the solutions can only be approximated *numerically*, with the help of computer algorithms. In some cases they can be expressed in algebraic form and then solved. We will only consider such cases here for two types of functions, exponential and logarithmic functions.

2.7.1 Exponential Equations

Exponential equations can be solved in two cases:

1) In case the equation consists of exponentials with different bases that are not added or subtracted, we can find a solution by applying a logarithm with any base to the equation.
2) In case exponentials of the same base k are added (e.g. 2^1, 2^2 and 2^0, that have the same base $k = 2$), we can substitute $y = k^x$ for an exponential equation in x.

Example 2.20

(Example of case 1) Solve the equation:

$$3^x = 4^{x-2} \cdot 2^x$$

In this case no exponentials are added or subtracted and we can apply a logarithm with any base to both sides of the equation. We will try the natural logarithm with base e, but it could be any base, and in fact base 2 might be more straightforward.

Remember that $\log_a x^b = b\log_a x$ and $\log_a (x \cdot y) = \log_a x + \log_a y$ (Sect. 1.2.2). If we now apply $\ln$ to both sides of the equation we get:

$$\ln 3^x = \ln \left(4^{x-2} \cdot 2^x\right) \quad \rightarrow \quad x \ln 3 = \ln 4^{x-2} + \ln 2^x \rightarrow$$

$$x \ln 3 = (x - 2) \ln 4 + x \ln 2$$

If we group over x we find that

$$x = \frac{2 \ln 4}{- \ln 3 + \ln 4 + \ln 2}$$

Example 2.21

(Example of case 2) Solve the equation:

$$3^{2x-1} = 9^x - 3$$

Remember that $3^2 = 9$, thus we can substitute $y = 3^x$. Since $3^{2x} = (3^x)^2 = y^2$, and $3^{2x-1} = \frac{3^{2x}}{3} = \frac{y^2}{3}$ we can rewrite the equation to

$$\frac{y^2}{3} = y^2 - 3$$

which we can then rewrite to $3y^2 - y^2 - 9 = 0$. Thus $2y^2 - 9 = 0$, and we know how to solve this by using e.g. the quadratic equation rule in Sect. 2.5.2.

Exercise

2.8. Find the roots of the following equations using logarithms or substitution:

a) $3^{6x} = 9$
b) $8 + 6^{2x+1} = 44$
c) $10e^{2x} - 30e^x + 15 = 0$

2.7.2 Logarithmic Equations

Logarithmic equations can be solved in the following two cases:

1) If the equation contains one or more logarithms of the same expression (say $P(x)$), then we can use the substitution:

$$y = \log_a P(x)$$

2) If the equation contains a linear combination of logarithms with the same base, we can use the arithmetic rules for logarithms (Sect. 1.2.2) to solve the equation.

Example 2.22

(Example of case 1). Solve the equation:

$$m\left(\ln\left(x^2 + 4\right)\right)^2 + n = a\sqrt{\left(\ln\left(x^2 + 4\right)\right)^2 + b}$$

According to the solution provided for case 1) we can rewrite

$$y = \ln\left(x^2 + 4\right)$$

(continued)

Example 2.22 (continued)

to get

$$my^2 + n = a\sqrt{y^2 + b}$$

If we square both sides, this results in

$$\left(my^2 + n\right)^2 = a^2 y^2 + a^2 b$$

which has now become a regular polynomial equation in y. After solving for y, x can be determined according to

$$x = \sqrt{e^y - 4}$$

Example 2.23

(Example of case 2). Solve the equation:

$$2\log_5(3x - 1) - \log_5(12x + 1) = 0$$

Rewriting this equation using the arithmetic rules for logarithms yields:

$$\log_5 \frac{(3x - 1)^2}{12x + 1} = \log_5 1$$

and thus

$$\frac{(3x - 1)^2}{12x + 1} = 1$$

This equation has two solutions ($x_1 = 0$ and $x_2 = 2$). However, if we now verify the solutions by substituting x_1 and x_2 in the original equation, the first logarithm has $3x_1 - 1 = -1$ as its argument, and $\log_5 - 1$ is not defined, whereas the second logarithm ($\log_5(12x_1 + 1) = \log_5 1 = 0$) is well defined, so we will discard the first solution. Therefore, the solution of the equation is only $x = 2$.

Exercise

2.9. Solve the following equations using substitution and logarithmic arithmetic:

    a) $\log_5 x - (\log_5 x)^2 = 0$
    b) $2\log_5(3x - 1) - \log_5(12x + 1) = 0$

2.8   Inequations

2.8.1 Introducing Inequations

Inequations are used in daily life almost as frequently as equations. They are used whenever we have to consider a lower or higher limit. For example, when driving, there is a maximum speed limit. If we take over a car that is driving slowly, are we going to have to drive too fast? If you spend 200 Euros on a new printer today, what is the maximum amount you can spend on food to not go over the daily withdrawal limit of your bank? For how many family members can you buy a plane ticket, before you hit the limit of your credit card? While equations include an equality (=) sign, inequations have an inequality symbol between two mathematical expressions as given in Table 2.1.

2.8.2 Solving Linear Inequations

As you need to know how to solve equations before you start solving any inequation, we started this chapter with an extensive introduction to equation solving. Inequations are solved using the same techniques as for equations, except that one needs to take care of the direction of the inequality symbol. The main rule is that if you multiply an inequation with a negative number you need to swap the direction of the inequality symbol, i.e. greater than becomes less than and vice versa.

For example, the inequation

$$-3x > 5$$

is solved by multiplying the whole inequation by $-\frac{1}{3}$. In this case the inequality symbol needs to change direction so the inequation becomes:

$$-\frac{1}{3}(-3x) < -\frac{1}{3} \cdot 5$$

Therefore, the solution is

$$x < -\frac{5}{3}$$

Table 2.1 Inequality symbols and their meanings

<	less than
>	greater than
≤	less than or equal to
≥	greater than or equal to
≠	not equal to

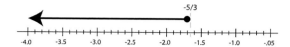

Fig. 2.2 Illustration of the set of solutions for the inequation $-3x > 5$. The *black arrow* represents all solutions. Any number smaller than $-5/3$ is a solution. The *black dot* indicates that the solution does not include the boundary value.

Table 2.2 Rules for changing the inequality symbol direction

Inequality symbol	Inequality symbol after multiplication by -1
$<$	$>$
$>$	$<$
$\leq$	$\geq$
$\geq$	$\leq$

Another important thing to remember is that the solution of an inequation is not a single value, as for linear equations, but a set of numbers that satisfies a certain criterion. I will explain that for the example above.

The solution of the above inequation is any x smaller than $-\frac{5}{3}$. This is illustrated in Fig. 2.2.

Table 2.2 summarizes what happens to inequality symbols in an inequation when both sides are multiplied by -1.

Inequations can also have three parts in the case where the solution of the inequation is limited both from above and from below. An example is given by:

$$-2 < 3x < 5$$

The solution of this inequation can be found by dividing all parts by 3. Since three is a positive number, the directions of all inequality symbols will remain the same:

$$-\frac{1}{3} \cdot 2 < x < \frac{1}{3} \cdot 5 \quad \rightarrow \quad -\frac{2}{3} < x < \frac{5}{3}$$

We can also write this as

$$x \in \left(-\frac{2}{3}, \frac{5}{3}\right)$$

Here, $\in$ stands for 'is an element of' or 'belongs to' and the ( ) brackets indicate that the boundary values are not included in the solution. Graphically this solution is illustrated in Fig. 2.3.

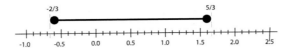

Fig. 2.3 Illustration of the set of solutions for the inequation $-2 < 3x < 5$. The *black dots* indicate that the solution does not include the boundary values.

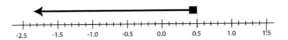

Fig. 2.4 Illustration of the set of solutions for the inequation $3 \geq 6x$. The *black square* indicates that the solution includes the boundary value.

Table 2.3 Examples of bracket use in linear inequations

Example	Same as
$x \in \left(-\frac{1}{2}, \frac{1}{2}\right)$	$-\frac{1}{2} < x < \frac{1}{2}$
$x \in \left[-\frac{1}{2}, \frac{1}{2}\right)$	$-\frac{1}{2} \leq x < \frac{1}{2}$
$x \in \left[\frac{-1}{2}, \frac{1}{2}\right]$	$-\frac{1}{2} \leq x \leq \frac{1}{2}$
$x \in \left(-\infty, \frac{1}{2}\right)$	$x < \frac{1}{2}$
$x \in \left(-\infty, \frac{1}{2}\right]$	$x \leq \frac{1}{2}$

But what about inequations like $3 \geq 6x$? This inequation includes the inequality symbol $\geq$, which stands for 'greater than or equal to' (See Table 2.1). If we apply already known rules for equation solving to solve this inequation we find that:

$$\frac{3}{6} \geq x \;\rightarrow\; \frac{1}{2} \geq x \;\rightarrow\; x \leq \frac{1}{2} \;\rightarrow\; x \in \left(-\infty, \frac{1}{2}\right]$$

This means that the solutions of the equation $3 \geq 6x$ are all numbers smaller than $\frac{1}{2}$ *including* $\frac{1}{2}$. The bracket] indicates that the boundary value is included in the solution. In line with the previous examples we can then illustrate this set of solutions as in Fig. 2.4.

Some further examples of the use of brackets to denote sets of numbers are provided in Table 2.3.

Exercise

2.10. Find the solution to the following inequations:

    a) $3x + 7 > 2x - 5$
    b) $3 - 5x < 7x - 2$
    c) $4 \leq 7x < 6$

2.8.3 Solving Quadratic Inequations

Quadratic inequations can be solved by using general knowledge about solving linear inequations while at the same time taking into account the quadratic nature of the inequation. Remember that the solution to a quadratic equation is determined by the roots of a parabola. To solve a quadratic inequation you thus need to:

1. Decide whether the parabola is concave up or concave down,
2. find the roots of that parabola and then
3. create the set of solutions.

This may sound pretty abstract but will become clear when discussing an example.

Example 2.24

Determine the solution of the inequation:

$$4x^2 - 9 < 0$$

First we realize that $4 > 0$, thus the parabola is concave up. That means that there are two solutions to the equation

$$4x^2 - 9 = 0$$

We will look at the part of the parabola that is <0, thus the part between the two solutions of the equation. Let's solve this equation by factoring (Sect. 2.5.3).

$$4x^2 - 9 = (2x - 3)(2x + 3)$$

Thus its roots are $x_1 = \frac{3}{2}$ and $x_2 = -\frac{3}{2}$. Here $x_2 < x_1$. Visually, we can now determine that the solution of the quadratic inequation is

$$x \in \left(\frac{-3}{2}, \frac{3}{2}\right)$$

as is also illustrated in the figure.

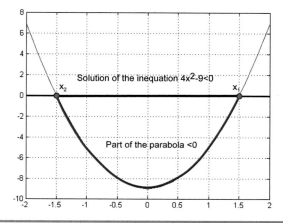

Example 2.25

Determine the solution of the inequation in Example 2.24, but then with the opposite inequality symbol:

$$4x^2 - 9 > 0$$

We can now immediately determine the solution with the help of the figure in Example 2.24:

$$x \in \left(-\infty, \frac{-3}{2}\right) \cup \left(\frac{3}{2}, \infty\right)$$

Here, $\cup$ denotes the *union* of two sets.

Exercise

2.11. Find the solution to the following inequations:

  a) $2x^2 + 7x + 5 \geq 0$
  b) $x^2 - 9 < 0$
  c) $x^2 - 3 > 5x + 2$

2.9   Scientific Example

Equivalent dose for anti-psychotic medication

My favorite example of equation use in neuroscience is based on a now widely used method to calculate the *equivalent dose* for *antipsychotics* (Andreasen et al. 2010). These are medications that can be used to suppress symptoms of *psychosis* and an equivalent dose is a dose which would offer an equal effect between different antipsychotics. The term equivalent dose is also used for other types of medications, such as e.g., *analgesics*. The importance of being able to calculate equivalent doses is that patients sometimes already use other types of antipsychotics (or analgesics) and doctors want to be able to track the total strength of the medication. Also, when comparing medication dosages between patients who use different medications, this is important. Here, we discuss a specific example.

Example 2.26

A patient is known to daily use 45 mg of Mirtazipine, 10 mg of Zyprexa and 15 mg of Abilify. How much is that expressed in mg of haloperidol?

Mirtazipine is an antidepressant, which will not count towards the equivalent dose of haloperidol. The other two drugs are antipsychotics where it should be known that Abilify is a brand of aripiprazole, and Zyprexa is based on olanzapine.

From the table in Fig. 2.5 we can derive that a dose of x mg of haloperidol is equivalent to

$$y_a = 4.343x^{0.645}$$

(continued)

Example 2.26 (continued)

Medications	Formulas and Equivalents			
	Formula[a] (x = CPZ)	Chlorpromazine Equivalent (mg)	Formula[b] (x = Haloperidol)	Haloperidol Equivalent (mg)
Atypical Antipsychotics				
Aripiprazole	$y = 0.255x^{0.700d}$	6.42	$y = 4.343x^{0.645d}$	6.79
Clozapine	$y = 2.027x^{0.863f}$	108	$y = 66.58x^{0.796e}$	115.61
Olanzapine	$y = 0.086x^{0.870e}$	4.75	$y = 2.900x^{0.805e}$	5.07
Quetiapine	$y = 2.806x^{0.852f}$	142	$y = 88.16x^{0.786e}$	151.97
Risperidone	$y = 0.019x^{0.924f}$	1.32	$y = 0.790x^{0.851e}$	1.43
Ziprasidone	$y = 2.805x^{0.628d}$	50.5	$y = 35.59x^{0.578d}$	53.13
Typical Antipsychotics				
Chlorpromazine	$y = x$	100	$y = 56.98x^{0.923f}$	108.04
Fluphenazine	$y = 0.011x^{1.112d}$	1.76	$y = 0.940x^{1.028d}$	1.92
Haloperidol	$y = 0.013x^{1.082f}$	1.84	$y = x$	2
Perphenazine	$y = 0.071x^{0.994f}$	6.90	$y = 3.937x^{0.919f}$	7.44
Thioridazine	$y = 0.989x^{0.973f}$	87.3	$y = 50.51x^{0.898f}$	94.14
Thiothixene	$y = 0.057x^{0.967f}$	4.91	$y = 2.852x^{0.892f}$	5.29
Trifluoperazine	$y = 0.066x^{0.939f}$	5.09	$y = 3.001x^{0.866f}$	5.47
Fluphenazine decanoate (mg/2–3 wk)	$y = 0.163x^{0.843c}$	7.91	$y = 4.921x^{0.778c}$	8.44
Haloperidol decanoate (mg/4 wk)	$y = 0.635x^{0.872d}$	35.3	$y = 21.662x^{0.803d}$	37.8

[a]Chlorpromazine is represented by "x" in this column, such that CPZ 100 mg would yield the equivalents in the next column over to the right.
[b]Haloperidol is represented by "x" in this column, such that haloperidol 2 mg would yield the equivalents in the next column over to the right.
[c]$R^2 > .96$.
[d]$R^2 > .97$.
[e]$R^2 > .98$.
[f]$R^2 > .99$.

Fig. 2.5 Formulas for calculating dose equivalents using regression with power transformation, and chlorpromazine and haloperidol equivalents based on them. Table reprinted from Andreasen et al. (2010). Antipsychotic dose equivalents and dose-years: a standardized method for comparing exposure to different drugs. Biol. Psychiatry 67, 255–262, with permission from Elsevier.

mg of Aripiprazole, and that a dose of x mg of haloperidol is equivalent to

$$y_o = 2.900x^{0.805}$$

mg of Olanzapine.

Thus, we want to determine x from the equations above, while we know the doses of aripiprazole ($y_a = 15$) and olanzapine ($y_o = 10$), as well as the full dosage per day (the sum of the two doses). You now know how to do this:

$$x_a = \left(\frac{y_a}{4.343}\right)^{\frac{1}{0.645}}$$

$$x_o = \left(\frac{y_o}{2.9}\right)^{\frac{1}{0.805}}$$

Here, x_a is the equivalent dose of haloperidol for the daily dose of aripiprazole and x_o is the equivalent dose of haloperidol for the daily dose of olanzapine. Note that $x = x_a + x_o$ is the daily dose of antipsychotics in haloperidol terms. Thus

$$x_a = \left(\frac{15}{4.343}\right)^{\frac{1}{0.645}} = 6.8324$$

$$x_o = \left(\frac{10}{2.9}\right)^{\frac{1}{0.805}} = 4.6540$$

and the daily equivalent dose of haloperidol (in mg) is

$$x = x_a + x_o = 11.4864$$

Thus this patients receives the equivalent of 11.49 mg of haloperidol daily.

Glossary

Analgesic medication to relief pain; painkiller

Algebraic using an approach in which only mathematical symbols and arithmetic operations are used

Antipsychotic medication used to treat psychosis

Arithmetic operations between numbers, such as addition, subtraction, multiplication and division

Concave hollow inward

Determinant a scalar calculated from a matrix; can be seen as a scaling factor when calculating the inverse of a matrix (see also Sect. 5.3.1)

Elimination eliminating an unknown by expressing it in terms of other unknowns

Equation a mathematical expression that states that two quantities are equal

Equivalent dose dose which would offer an equal effect between different medications

Function a mathematical relation, like a recipe, describing how to get from an input to an output

Independent here: equations that cannot be transformed into each other by multiplication

Least common denominator the least number that is a multiple of all denominators

Linear a function or mathematical relationship that can be represented by a straight line in 2D and a plane in 3D; can be thought of as 'straight'

Nonlinear not linear

Numerically (solving) to find an approximate answer to a mathematical problem using computer algorithms

Polynomial an expression consisting of a sum of products of different variables raised to different non-negative integer powers

Psychosis a mental condition that can have many different symptoms including hallucinations

Rational equation equation that has a rational expression on one or both sides in which the unknown variable is in one or more of the denominators

Root solution of a polynomial equation

Substitution replacing a symbol or variable by another mathematical expression

Transcendental a number that is not the root of a polynomial with integer coefficients; most well-known are e and π

Union the union of two sets is the set that contains all elements in both sets

Unique here: a single solution to an equation

Unknown variable in an equation for which the equation has to be solved; an equation can have multiple unknowns

Variable alphabetic character representing a number

Vertex peak of a parabola

Y-intercept intercept of a curve with the y-axis

Symbols Used in This Chapter (in Order of Their Appearance)

$=$	equal to
$\rightarrow$	implies
$\neq$	not equal to
$\mathbb{C}$	complex numbers
log	logarithm
ln	natural logarithm (base e)
i	complex unity

$<$	less than
$>$	greater than
$\leq$	less than or equal to
$\geq$	greater than or equal to
$\neq$	not equal to
$x \in \left(-\frac{1}{2}, \frac{1}{2}\right)$	$-\frac{1}{2} < x < \frac{1}{2}$
$x \in \left[-\frac{1}{2}, \frac{1}{2}\right)$	$-\frac{1}{2} \leq x < \frac{1}{2}$
$x \in \left[-\frac{1}{2}, \frac{1}{2}\right]$	$-\frac{1}{2} \leq x \leq \frac{1}{2}$
∞	infinity
$\cup$	unification

Overview of Equations for Easy Reference

General form of linear equation
Any linear equation with one unknown x can be written as $ax = b$ where a and b are constants.

Arithmetic rules useful for solving linear equations
If $a = b$ and $c \in \mathbb{C}$ then

$$a + c = b + c$$
$$ac = bc$$

If $a = b$, $c \in \mathbb{C}$ and $c \neq 0$ then

$$\frac{a}{c} = \frac{b}{c}$$

If F is any function and $a = b$ then $F(a) = F(b)$

$$F(a) = F(b)$$

Cramer's rule for a system of 2 linear equations
The system of two equations with two unknowns x and y

$$ax + by = c$$

$$dx + ey = f$$

has the solution

$$x = (ce - bf)/(ae - bd)$$

$$y = (af - cd)/(ae - bd)$$

when

$$ae - bd \neq 0.$$

Quadratic equation rule
The solution of

$$ax^2 + bx + c = 0$$

is given by

$$x_{1,2} = \frac{-b \pm \sqrt{b^2 - 4ac}}{2a}$$

Factor multiplication rule
General rule:

$$(x + y)(x + z) = x^2 + (y + z)x + yz$$

Special cases:

$$(x + y)^2 = x^2 + 2xy + y^2$$

$$(x - y)^2 = x^2 - 2xy + y^2$$

$$x^2 - y^2 = (x - y)(x + y)$$

Rules for changing the direction of the inequality symbol

Inequality symbol	Inequality symbol after multiplication by -1
$<$	$>$
$>$	$<$
$\leq$	$\geq$
$\geq$	$\leq$

Answers to Exercises

2.1. Suppose each of the granddaughters inherits x coins. Then we can write $3x + \frac{1}{2} \cdot 60 = 60 \rightarrow 3x = 60 - 30 \rightarrow 3x = 30 \rightarrow x = 10$. Thus each granddaughter will inherit ten rare coins.

2.2. a. $x = -8$
   b. $x = -23$
   c. $3x - 9 = 33 \rightarrow 3x = 33 + 9 \rightarrow x = \frac{42}{3} \rightarrow x = 14$
   d. $x = \frac{65}{5} = 13$
   e. $4x - 6 = 6x \rightarrow (4 - 6)x - 6 = 0 \rightarrow -2x = 6 \rightarrow x = -3$
   f. $8x - 1 = 23 - 4x \rightarrow 12x = 24 \rightarrow x = 2$

2.3. a) $x = 0$

  b) $x = \frac{-9}{13}$

2.4. a) $x = 4; y = 0$

  b) $x = 1; y = -\frac{3}{2}$

  c) $x = \frac{3}{2}; y = -1$

  d) $x = 3; y = 0$

2.5. a) $x = 2; y = -3$

  b) $x = 1; y = 2$

  c) $x = 10; y = 3$

  d) $x = 2; y = -1; z = 1$

2.6. a) $x_1 = -2\frac{1}{2}$ and $x_2 = -1$

  b) $x_1 = -3$ and $x_2 = 3$

  c) $x_1 = \frac{5-3\sqrt{5}}{2}$ and $x_2 = \frac{5+3\sqrt{5}}{2}$

  d) $x_1 = \frac{5-\sqrt{37}}{2}$ and $x_2 = \frac{5+\sqrt{37}}{2}$

  e) $x_1 = 3 - \frac{1}{2}\sqrt{42}$ and $x_2 = 3 + \frac{1}{2}\sqrt{42}$

  f) $x_1 = -3 - \frac{3}{2}\sqrt{3}$ and $x_2 = -3 + \frac{3}{2}\sqrt{3}$

2.7. a) $(x-5)(x-2)=0$ thus $x_1=2$ and $x_2=5$

  b) $(x-4)(x-1)=0$ thus $x_1=1$ and $x_2=4$

  c) $(x+4)(x-2)=0$ thus $x_1=-4$ and $x_2=2$

  d) We can rewrite this equation to $2x(x-5)(x-2)=0$ thus $x_1=0$, $x_2=2$ and $x_3=4$

2.8. a) Using that $9=3^2$, we have to solve $6x=2$ and thus $x = \frac{1}{3}$.

  b) We can rewrite $6^{2x+1}=36=6^2$ so that we have to solve $2x+1=2$ and thus $x = \frac{1}{2}$.

  c) Substitute $y=e^x$, so that $y^2=e^{2x}$, resulting in the quadratic equation $10y^2-30y+15=0$, which has solutions $y_1 = \frac{3}{2}+\frac{1}{2}\sqrt{3}$ and $y_2 = \frac{3}{2}-\frac{1}{2}\sqrt{3}$. Then $x_1 = \ln\left(\frac{3}{2}+\frac{1}{2}\sqrt{3}\right)$ and $x_2 = \ln\left(\frac{3}{2}-\frac{1}{2}\sqrt{3}\right)$.

2.9. a) First we substitute $y=\log_5 x$. Then the equation can be rewritten as $y-y^2=0$, which has two solutions $y_1 = 0$ and $y_2 = 1$. Hence $\log_5 x_1=0$ and $\log_5 x_2=1$ and thus $x_1=1$ and $x_2=5$

  b) We can rewrite this equation using the rules for logarithms to $\log_5\frac{(3x-1)^2}{12x+1} = \log_5 1$, so that we have to solve the equation $\frac{(3x-1)^2}{12x+1} = 1$, which has solutions $x_1=0$, $x_2=2$.

2.10. a) $x \in (-12, \infty)$

  b) $x \in \left(\frac{5}{12}, \infty\right)$

  c) $x \in \left[\frac{4}{7}, \frac{6}{7}\right)$

2.11. a) We first solve the equation $2x^2 + 7x + 5 = 0$ which has two solutions $x_1 = \frac{-5}{2}$ and $x_2 = -1$. Because the coefficient of x^2 is larger than zero, $a > 0$, the parabola has a minimum and is concave up. We are looking for those solutions where the curve is ≥ 0. Thus

$$x \in \left(-\infty, \frac{-5}{2}\right] \cup [-1, \infty)$$

b) In a similar approach as for Exercise 2.11a we find that:

$$x \in (-3, 3)$$

c) We first rewrite the inequation to $x^2 - 5x - 5 > 0$. Its related equation has two solutions $x_{1,2} = \frac{5 \pm \sqrt{45}}{2}$. Thus

$$x \in \left(-\infty, \frac{5 - 3\sqrt{5}}{2}\right) \cup \left(\frac{5 + 3\sqrt{5}}{2}, \infty\right)$$

References

Online Sources of Information

https://en.wikipedia.org/wiki/Linear_equation
http://www.mathsisfun.com/algebra/quadratic-equation-graph.html
http://www.mathsisfun.com/geometry/parabola.html
https://www.khanacademy.org/

Books

Bronstein, Semendjajev, *Taschenbuch der Mathematik [Handbook of Mathematics]* (Teubner, Leipzig, 1984)

Papers

N.C. Andreasen, M. Pressler, P. Nopoulos, D. Miller, B.C. Ho, Antipsychotic dose equivalents and dose-years: a standardized method for comparing exposure to different drugs. Biol. Psychiatr. **67**, 255–262 (2010)

3

Trigonometry

Natasha Maurits

After reading this chapter you know:

- what the three main trigonometric ratios are and how they can be calculated,
- how to express angles in degrees and radians,
- how to sketch the three main trigonometric functions,
- how to sketch functions of the form $A\sin(Bx+C)+D$ and how to interpret A, B, C and D,
- what Fourier (or spectral or harmonic) analysis is and
- how Fourier analysis is used in real-life examples.

3.1 What Is Trigonometry and How Is It Applied?

The word trigonometry is derived from the Greek words trigon (triangle) and metron (measure) and thus literally means 'measuring triangles'. In particular, trigonometry cleverly employs the relation between sides and angles in triangles. Even though many people find trigonometry a rather abstract branch of mathematics, it actually developed from very practical applications in astronomy and land measurement, where distances from one point to another have to be obtained. Let me give you an example of how measuring triangles and employing the relation between sides and angles in a triangle can help obtain the distance between two points that are too far apart to be measured with a tape measure. Suppose you want to know the height of the beautiful Martini Tower, a landmark of our home city of Groningen, to determine whether it is taller than the Dom Tower in Utrecht. To do this, you only need a tape measure to measure (1) the distance between you and an object between you

N. Maurits (✉)
Department of Neurology, University Medical Center Groningen, Groningen, The Netherlands
e-mail: n.m.maurits@umcg.nl

© Springer International Publishing AG 2017
N. Maurits, B. Ćurčić-Blake, *Math for Scientists*, DOI 10.1007/978-3-319-57354-0_3

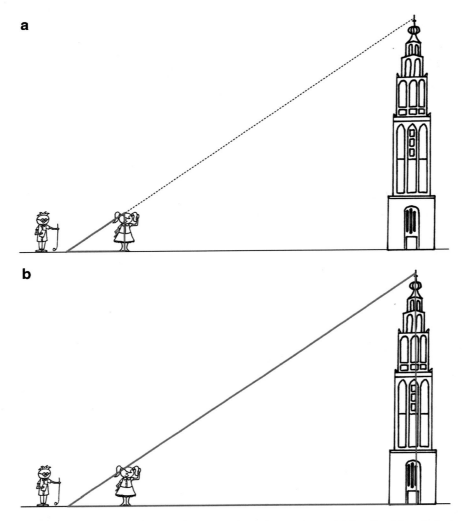

Fig. 3.1 Measuring the height of Martini Tower using trigonometry. **(a)** Smaller triangle between you and object. **(b)** Larger similar triangle between you and Martini Tower.

and the Martini Tower, (2) the height of this object and (3) the distance between you and the Martini Tower (see Fig. 3.1).

The smaller triangle in Fig. 3.1a and the larger triangle in Fig. 3.1b are *similar*: they have the same angles, so the ratio between the height of the Martini Tower and the distance between you and the Martini Tower (142.40 m) and the ratio between the length of the object (a helpful bystander who is 1.70 m tall) and the distance between you and the object (2.50 m) must be the same. The height of the Martini Tower can then be calculated from the equality height/142.40 = 1.70/2.50, giving height = 142.40 × 1.70/2.50 = 96.83 m.

A similar procedure can be used to calculate the distance from a boat to a point on the shore, such as a lighthouse, by cleverly choosing the triangles. In this case, you would construct two different triangles, the first with two of its vertices on the boat—close to each other so that you can measure the distance in between—and the third at the lighthouse

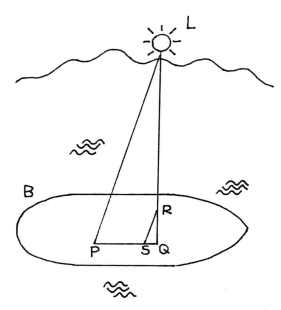

Fig. 3.2 Measuring the distance to a lighthouse *L* on the shore from a boat *B* using trigonometry. *PQL*: reference triangle. *SQR*: similar triangle. Here: PQ:SQ = QL:QR. Since PQ, SQ and QR can be measured, QL can be derived.

L and another one with the same shape (angles), but smaller, with the third vertex also on the boat close to you (Fig. 3.2).

In general, trigonometry allows calculating the length of an edge of a *right-angled triangle* if an angle and the length of another edge are known, or an angle if the lengths of two other edges are known, using similar triangles. With this approach, it is also possible to calculate, for example, distances to a star from earth or between celestial bodies. More developed forms of trigonometry are used in architecture, medicine and biology, music and acoustics, neuroimaging and astronomy, and I will show you how in some examples later in this chapter.

3.2   Trigonometric Ratios and Angles

After a while it became convenient to refer to the ratios between the lengths of two edges in a right-angled triangle with specific names. As these ratios do not depend on the size of the triangle, but only on the angles, tables were drafted with the values of these ratios for different angles. The three main trigonometric ratios were named *sine*, *cosine* and *tangent*.

For the angle α in Fig. 3.3 the sine (S) is defined as the ratio between the opposite edge (O) and the *hypothenuse* (H), the cosine (C) as the ratio between the adjacent (A) edge and the hypotenuse and the tangent (T) as the ratio between the opposite edge and the adjacent edge. To summarize:

$$sine = O/H$$
$$cosine = A/H$$
$$tangent = O/A = sine/cosine$$

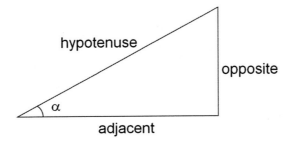

Fig. 3.3 Basic triangle to define trigonometric ratios.

To memorize these trigonometric ratios a mnemonic for the acronym SOHCAHTOA—Sine (Opposite over Hypotenuse), Cosine (Adjacent over Hypotenuse), Tangent (Opposite over Adjacent)—can be useful (see for many examples: http://acronyms.thefreedictionary.com/Sohcahtoa).

Exercise

3.1. Find the values of the three main trigonometric ratios (sine, cosine, tangent) for each of the following angles θ:

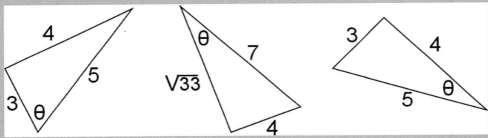

3.2. Suppose you are building an ellipsoid dinner table for which you would like to use four (square) legs on the positions indicated in the figure (table seen from below, arrows point to the center of the legs). The legs are connected by trapezoid beams that you now need to saw. The machine you use can saw angles, but what are the angles 1, 2, 3 and 4?

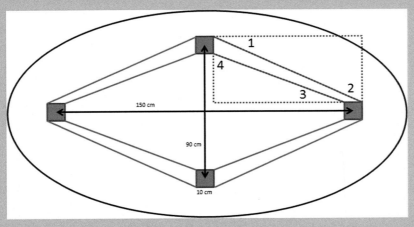

Employing Pythagoras' theorem, we can easily derive the first of a number of useful relationships between the trigonometric ratios. Just to remind you, Pythagoras' theorem states that the sum of the squares of the edges adjacent to a right angle in a triangle is equal to the square of the hypothenuse (or in Fig. 3.3: Opposite^2 + Adjacent^2 = Hypothenuse^2). Thus:

$$\sin^2 + \cos^2 = \left(\frac{O}{H}\right)^2 + \left(\frac{A}{H}\right)^2 = \frac{O^2 + A^2}{H^2} = \frac{H^2}{H^2} = 1 \qquad (3.1)$$

This relationship between sine and cosine is always true, independent of the angle, and allows easily deriving the value of the cosine if the value of the sine is known and vice versa. This relationship also tells us that if we would plot cosine versus sine for all possible values of the angle, a circle with radius one would result (remember that the definition of a *unit circle* in the (x,y)-plane is $x^2 + y^2 = 1$). This knowledge will be very useful when exploring the trigonometric ratios as trigonometric *functions* a little later in this chapter. In addition, when defining the trigonometric ratios on the unit circle, life gets a bit easier as the hypothenuse (the radius of the unit circle) becomes 1 and drops out of the ratios for sine and cosine (Fig. 3.4).

The edge lengths of the smaller triangle in Fig. 3.4a are thus equal to the sine and cosine. To understand why the tangent is represented by the green line in Fig. 3.4a requires a little more thinking. Remember that the ratios only depend on the angle and not on the size of the triangle; thus, they are the same in similar triangles. As the tangent is the ratio between the sine and the cosine (the length of the red edge divided by the length of the blue edge) and for the larger similar triangle in Fig. 3.4a the adjacent edge has length 1, the tangent is here simply equal to the length of the green line. Also, as the trigonometric ratios in a triangle are only defined for angles between 0 and 90 degrees (the angles can simply not get larger within

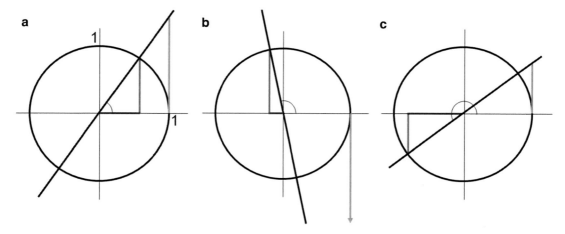

Fig. 3.4 Trigonometric ratios on the unit circle (**a**) for angles between 0 and 90 degrees, (**b**) for angles between 90 and 180 degrees, (**c**) for angles between 180 and 270 degrees. The sine is equal to the length of the *red line*, the cosine to the length of the blue line and the tangent to the length of the *green line*.

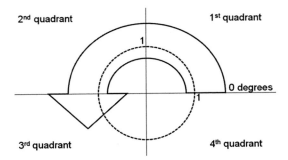

Fig. 3.5 Illustration of the definition of angles and quadrants employing a circle.

a right-angled triangle), the representation on the unit circle allows extending these definitions to larger angles, making use of the periodicity of the trigonometric functions. Before doing that, we need to define angles and quadrants on the circle (Fig. 3.5).

By definition, angles are defined starting from the positive x-axis going counter-clockwise. Thus, for example, a 90-degree (or 90°, as the symbol for degrees is °) angle is defined between the positive x-axis and the positive y-axis, while a 290° angle is defined between the positive x-axis and the negative y-axis. The first quadrant is then the top right one, the second the top left, the third the bottom left and the fourth the bottom right. When traversing from the fourth to the first quadrant one can either further increase the angle or start counting from 0° again, e.g. an angle of 390° is equal to an angle of 30°. Angles are thus given *modulo* 360°. Similarly, an angle of −30° is equal to an angle of 330°.

Now consider Fig. 3.4b. Here, the angle is larger than 90°, but the sine and cosine can still be derived from a (now different) triangle. Note that the cosine is now negative (even though length is always positive, the cosine edge of the triangle extends along the negative x-axis in the plot), making the tangent negative, as well, which is why we consider the downward edge of the similar triangle on the other side of the origin to derive the tangent. For angles between 180 and 270° a similar approach can be used (Fig. 3.4c). Note that sine and cosine are both negative for this quadrant, making the tangent positive again.

3.2.1 Degrees and Radians

In the previous text I already talked about angles, also assuming that you are familiar with right angles being 90°. If we go around the full circle in Fig. 3.5, we will have travelled 360°. You may wonder why this is not 100°, in line with the base 10 system we use for counting. This dates all the way back to the Babylonian astronomers of the last centuries BC who used a base 60 system to calculate (Neugebauer, 1969). Angles can also be expressed in other units, however. The *SI* unit for angle is the radian (or rad), which is more convenient to use than °. For example, that the derivative of the sine is the cosine, and that $\lim\limits_{x \to 0} \frac{\sin(x)}{x} = 1$ (see Chap. 6) only hold when using radians. Using radians, a whole revolution of a circle is compared to the circumference of a circle, i.e. $2\pi r$, where r is the radius of the circle. An angle in radians is the length of the corresponding arc on a unit circle (Fig. 3.6).

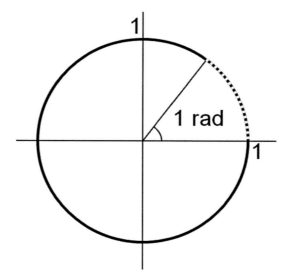

Fig. 3.6 Definition of an angle in radians: the dashed arc has a length of 1.

Note that in a circle with a radius that is twice as large the corresponding arc will also be twice as long. As angles do not change depending on the distance to the arc, in general radians measure the arc formed by the angle divided by the radius of that arc. For a full circle (360°), the arc is equal to the circumference of the circle $2\pi r$. Hence, when dividing by the radius of the circle r, we find that 360° equals 2π radians, or $1° = 2\pi/360 = \pi/180$ rad. Vice versa, $1 \text{ rad} = 360/2\pi° = 180/\pi°$. As angles are given modulo 360°, similarly they are given modulo 2π rad.

Exercises

3.3. Convert the following angles from degrees to radians:

(a) 30°
(b) 45°
(c) 60°
(d) 80°
(e) 90°
(f) 123°
(g) 260°
(h) −16°
(i) −738°

3.4. Convert the following angles from radians to degrees

(a) $2\pi/3$ rad
(b) $\pi/4$ rad
(c) $9\pi/4$ rad
(d) 0.763π rad
(e) π rad
(f) $\theta\pi$ rad
(g) θ rad

3.3   Trigonometric Functions and Their Complex Definitions

So far, we have only considered trigonometric ratios; sine, cosine and tangent were defined employing the lengths of the edges of a triangle. In other words, we can now derive the sine, cosine and tangent for every angle by considering an appropriate triangle as we did e.g. in the context of the unit circle in Fig. 3.4. However, being able to define the sine, cosine and tangent for every angle, implies that we can also consider these trigonometric ratios as functions of an angle. A function is like a recipe, describing how to get from an input to an output. For example, a function f that relates a value of x to its square root $\sqrt{x}$ could be given by $f(x) = \sqrt{x}$. Similarly, we can define functions that relate an angle α to its sine ($f(\alpha)$=sin (α)), cosine ($f(\alpha)$=cos(α)) or tangent ($f(\alpha)$=tan(α)). Note that in these functions we use abbreviations for the trigonometric ratios. Now that we have trigonometric functions we can also graph these functions in a coordinate system. Usually, when plotting any function, the input x is plotted along the horizontal x-axis and the output, the value of the function $f(x)$, along the y-axis.

A plot of sin(x) could be made manually by first making a table for pairs of (x, sin(x)), then plotting the points in the table in the (x,y)-plane and sketching a smooth curve through these points. The values for sin(x) could be obtained from a calculator. However, the fact that sin(x) is defined as the opposite edge in the triangle in the unit circle in Fig. 3.4 also helps in plotting (Fig. 3.7).

Imagine a point travelling along the unit circle, starting at the positive x-axis where the angle is 0°. This position corresponds to the origin in the plot of the sine function. Then, while the point is travelling along the circle, plot its varying distance from the x-axis for the corresponding angles in the plot of the sine function (fat dotted line in Fig. 3.7). When the point has travelled the full circle, a plot for sin(x) between 0 and 2π radians results (drawn line in Fig. 3.7, right). For the cosine function a similar approach could be taken, but now the points with varying distance from the y-axis are plotted for the corresponding angles (dotted

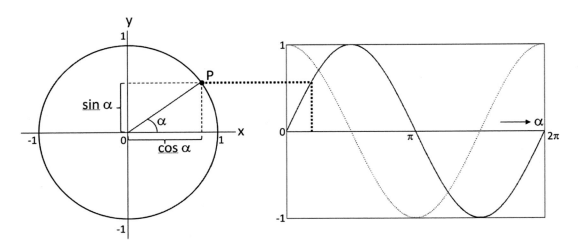

Fig. 3.7 Plotting sin(x) (right, *drawn line*) and cos(x) (right, *dotted line*) employing the unit circle on the left. The *fat dotted line* links the point P on the unit circle to the related point on the sine graph.

line in Fig. 3.7, right). Both the sine and cosine functions extend to the left and right, although these extensions are not drawn, as the point could continue travelling along the unit circle and also in the opposite direction (for negative angles). The trigonometric functions thus repeat themselves regularly (every 2π for the sine and cosine and every π for the tangent) and are therefore called *periodic*.

To obtain values for trigonometric functions for any angle it is usually easiest to use a calculator. However, there are a few values for $\sin(x)$ for specific angles x that are useful to remember (Table 3.1).

Exercises

3.5. Determine whether each of the following statements is true or false.

(a) $\cos(0°) = 0$
(b) $\sin(30°) = \frac{1}{2}\sqrt{3}$
(c) $\sin(45°) = 1/2$
(d) $\cos(45°) = \frac{1}{2}\sqrt{2}$
(e) $\cos(60°) = 1$

3.6. Find the value of x.

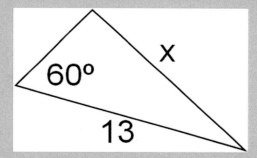

3.7. Sketch graphs of the following functions and provide exact values for the angles 0, 30, 45, 60 and 90 degrees.

(a) $\sin(x)$
(b) $\cos(x)$
(c) $\tan(x)$

The good thing about remembering the values in Table 3.1 is that – given that you are able to sketch the trigonometric functions - you can derive the values for the trigonometric functions for a whole bunch of other angles without having to remember them. Let me show you how with the help of Table 3.1 and Fig. 3.8.

Note first that $\cos(x)$ is simply shifted with respect to $\sin(x)$ by $\pi/2$ radians and that there is ample symmetry within and between the trigonometric functions. Regarding the

Table 3.1 Values for $\sin(x)$ to remember

x in rad	$\pi/6$	$\pi/4$	$\pi/3$	$\pi/2$	π
$\sin(x)$	0.5	$\frac{1}{2}\sqrt{2}$	$\frac{1}{2}\sqrt{3}$	1	0

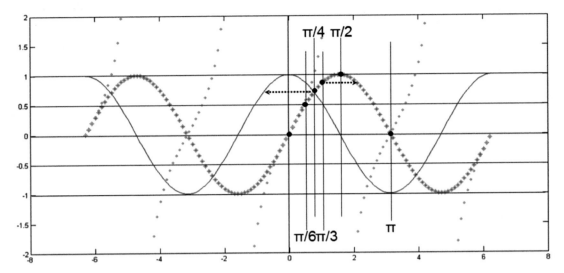

Fig. 3.8 Deriving other values for trigonometric functions employing the values in Table 3.1 (*black dots*). *Red*: sin(x), *blue*: cos(x)*green*: tan(x).

latter let's start with the cosine: this function remains the same when mirrored in the y-axis. Or expressed mathematically: $\cos(-x)=\cos(x)$. For the sine, a similar but slightly different relationship holds: this function is unchanged when rotated around the origin over 180°. Or mathematically: $\sin(-x)=-\sin(x)$. Finally, the tangent function behaves similarly as the sine function in terms of symmetry: $\tan(-x) = \frac{\sin(-x)}{\cos(-x)} = \frac{-\sin(x)}{\cos(x)} = -\tan(x)$.

To see how other values for trigonometric functions can now easily be obtained I give two examples in Fig. 3.8. The dashed arrow pointing to the left shows that (in radians) $\cos\left(-\frac{\pi}{4}\right) = \cos\left(\frac{\pi}{4}\right) = \sin\left(\frac{\pi}{4}\right) = \frac{1}{2}\sqrt{2}$. The dashed arrow pointing to the right illustrates that $\sin\left(\frac{2\pi}{3}\right) = \sin\left(\frac{\pi}{3}\right) = \frac{1}{2}\sqrt{3}$.

Exercise

3.8. Use the even-odd identities to simplify the expressions (i.e. get the minuses out of the brackets).

(a) $\cos(-2x)$

(b) $\tan\left(-\frac{\pi}{4}\right)$

(c) $\sin\left(-\frac{4\pi}{3}\right)$

In Exercise 3.8, you encountered the expression $\cos(-2x)$; I here now explain how this function is related to the basic function $\cos(x)$ and more generally how nonstandard trigonometric functions (and their graphs) are related to the standard trigonometric functions $\sin(x)$, $\cos(x)$ and $\tan(x)$. If you understand these general relationships, this will help you understand the widely applied Fourier or harmonic analysis that is explained later in this chapter.

It is easiest to think about this in a visual way: graphs for nonstandard trigonometric functions can be obtained by scaling and moving the standard curve. Mathematically this is expressed by choosing different values for the parameters A, B, C and D in the basic function:

$$A \sin (Bx + C) + D \tag{3.2}$$

Here A determines the amplitude (the largest and smallest values), B the period (how long it takes the function to repeat itself), C the phase (the shift in angle with respect to the basis function) and D the offset (the mean value around which the function oscillates). For example, the cosine function is the same as the sine function shifted by $\frac{\pi}{2}$ to the left, hence $\cos (x) = \sin \left(x + \frac{\pi}{2}\right)$. In many examples, a sine function approximates a variable that changes periodically over time, such as the tides, predator/prey populations or sleep-wake rhythms. In these cases the period, referred to by T, is given in seconds. The period also determines the frequency f of the sine as its inverse $f = 1/T$ (given in Hz $= 1/s$). The higher the frequency of the sine, the faster it repeats itself over time. Thus:

$\sin(x)$	has period	2π	and frequency	$\frac{1}{2\pi}$
$\sin(3x)$	has period	$\frac{2\pi}{3}$	and frequency	$\frac{3}{2\pi}$
$\sin(2\pi t)$	has period	1	and frequency	1
$\sin(4\pi t)$	has period	$\frac{1}{2}$	and frequency	2
$\sin(6\pi t + 50)$	has period	$\frac{1}{3}$	and frequency	3

Exercises

3.9. Which of the graphs below represents $\sin(x)$, $\sin(2x)$, $\sin(x)+2$, $\sin(x+2)$ and $2\sin(x)$?

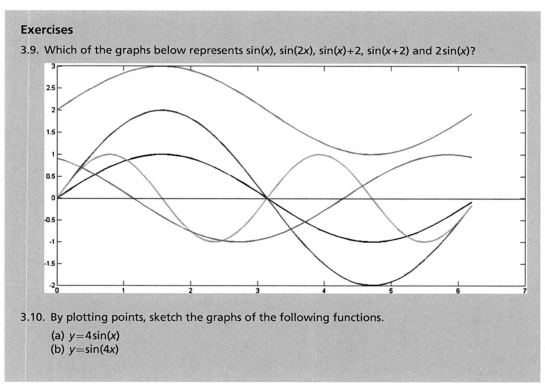

3.10. By plotting points, sketch the graphs of the following functions.
  (a) $y = 4\sin(x)$
  (b) $y = \sin(4x)$

(continued)

3.11. By plotting points, sketch the graphs of the following functions.
  (a) $y=2+\sin(x)$
  (b) $y=-\cos(x)$

3.12. What transformations should be used to obtain the following functions from the standard sine function?
  (a) $y=-3\sin(x)-5$
  (b) $y=2\sin(3x)$
  (c) $y=2\sin(x+\pi)+2$
  (d) $y=3\sin\left(2x-\frac{\pi}{2}\right)-1$
  (e) $y=-4\sin\left(\frac{\pi x}{5}\right)$
  (f) $y=-x+3\sin(2x-\pi)$

3.3.1 Euler's Formula and Trigonometric Formulas

There is an interesting relationship between the complex exponential function and the trigonometric functions known as Euler's formula:

$$e^{ix} = \cos(x) + i\sin(x) \tag{3.3}$$

It employs that a point in the complex plane can be described in polar (left hand of Eq. 3.3) as well as in Cartesian (right hand of Eq. 3.3) coordinates (see also Fig. 3.9 and Sect. 1.2.4).

The nice thing about Euler's formula is that it allows deriving many relationships between the trigonometric functions employing the relative ease of manipulating exponentials, compared to trigonometric functions. For example, starting with the following identity for exponentials:

$$e^{i(x+y)} = e^{ix}e^{iy},$$

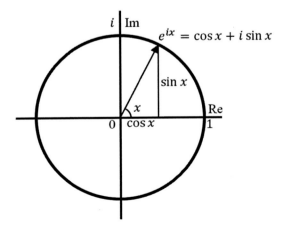

Fig. 3.9 The point cos(x)+isin(x) on the unit circle in the complex plane.

we can fill in Euler's formula on both sides to arrive at:

$$\cos(x+y) + i\sin(x+y) = (\cos(x) + i\sin(x))(\cos(y) + i\sin(y))$$

Expanding the right hand side (and using that $i^2 = -1$) we get:

$$\cos(x+y) + i\sin(x+y) = \cos(x)\cos(y) + i\cos(x)\sin(y) + i\sin(x)\cos(y) - \sin(x)\sin(y)$$

When we now equate the real and imaginary parts of this equation we find that:

$$\cos(x+y) = \cos(x)\cos(y) - \sin(x)\sin(y)$$
$$\sin(x+y) = \cos(x)\sin(y) + \sin(x)\cos(y)$$

Similarly (you could try to derive these equalities as an additional exercise):

$$\cos(x-y) = \cos(x)\cos(y) + \sin(x)\sin(y)$$
$$\sin(x-y) = \sin(x)\cos(y) - \cos(x)\sin(y)$$

and, employing Eq. 3.3:

$$\sin(2x) = \sin(x+x) = \cos(x)\sin(x) + \sin(x)\cos(x) = 2\sin(x)\cos(x)$$
$$\cos(2x) = \cos(x+x) = \cos(x)\cos(x) - \sin(x)\sin(x) = \cos^2(x) - \sin^2(x)$$

From the last equation, applying Eq. 3.1, we finally derive:

$$\cos(2x) = \cos^2(x) - \sin^2(x) = 2\cos^2(x) - 1 = 1 - 2\sin^2(x)$$

As touched upon in Sect. 1.2.4.2, the polar form of complex numbers also allows to e.g., find all 3 complex roots of the equation $z^3 = 1$ and not just the one obvious real root $z = 1$. To solve $z^3 = 1$ using complex numbers, we first substitute 1 by its polar form, by realizing that 1 is located on the real axis in the complex plane. It therefore has modulus 1 and the angle between the positive real axis and 1 is zero. Hence:

$$1 = 1 \cdot e^{0i}$$

Of course this is rather trivial. However, the angle φ is only defined modulo 2π, meaning that the angle could also be 2π or 4π or any other multiple of π. Thus:

$$1 = 1 \cdot e^{n \cdot 2\pi i} \text{ for } n = 0, 1, 2, \ldots$$

If we now look at the equation we are trying to solve (and use some of the arithmetic rules for exponents), we find that we are looking for a complex number z such that

$$z^3 = \left(re^{i\phi}\right)^3 = r^3 e^{3i\phi} = 1 \cdot e^{n \cdot 2\pi i}$$

Thus, we are looking for a real r such that $r^3 = 1$ and real φ such that $3i\,\varphi = 2n\pi i$, $n = 0$, 1, 2, ... A solution is thus $r = 1$ and $\phi = \frac{2}{3}n\pi$, $n = 0, 1, 2, \ldots$ or, in other words, the three complex roots are $z = 1$, $z = e^{\frac{2}{3}\pi i}$ and $z = e^{\frac{4}{3}\pi i}$. Note that these three complex roots are evenly distributed over de unit circle in the complex plane. In general, finding all n complex roots of the equation $z^n = 1$ works in exactly the same way. One of the roots is 1 and the other n-1 roots are evenly distributed over the unit circle in the complex plane.

3.4   Fourier Analysis

Trigonometry is probably most often used—across many, many scientific fields—in the form of Fourier analysis, which is also known as spectral or harmonic analysis. Harmonic analysis is a more general field of mathematics, though. In Fourier analysis, a function is decomposed into a sum of trigonometric functions (which can be seen as oscillatory components). The inverse procedure also exists and is known as inverse Fourier analysis: in that case a function is synthesized from its components (Fig. 3.10).

The procedure got its name from Jean Baptiste Joseph Fourier, who, in the early nineteenth century, was the first to realize that any periodic function could be decomposed into an infinite sum (or series) of trigonometric functions. Before Fourier, predecessors of Fourier analysis were already used to represent planetary and asteroid orbits. As is the case for many branches of mathematics, Fourier analysis was thus developed to solve practical problems, in this case for example in astronomy.

The frequencies in a Fourier series are not random, but are rather determined by the period T of the function that is considered. The lowest frequency in the series is 1/T and higher frequencies are simply multiples of this lowest frequency: i.e. 2/T, 3/T, 4/T etc. The lowest frequency is called the base, ground or fundamental frequency and the multiples are called harmonic frequencies. This is why this type of analysis is sometimes also referred to as harmonic analysis. Thus, when a signal with a period of 10 seconds is expressed as a Fourier series, the base frequency is $1/10 = 0.1$ Hz and the harmonic frequencies are 0.2, 0.3, 0.4 Hz etc. This does however not imply that each of these frequencies is actually present in the series (Fig. 3.11).

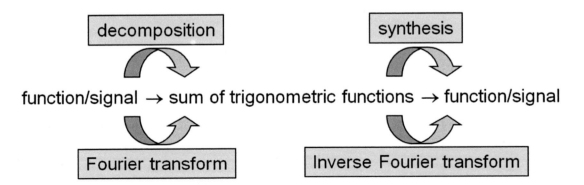

Fig. 3.10   Illustration of the bi-directional procedure of Fourier analysis and its inverse.

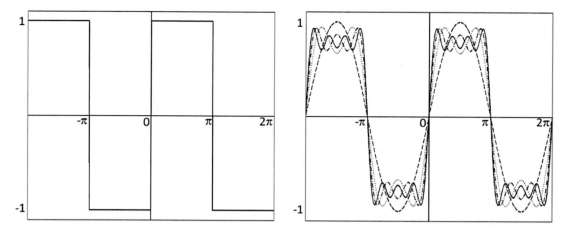

Fig. 3.11 *Left*: Example of a Fourier series expansion for a periodic block function (period 2π). *Right*: The Fourier series approximation of the signal is displayed when more and more terms are added. *Dashed line*: first approximation (sin(x)), *dotted line*: second approximation (sin(x) + 1/3 sin(3x)), *dash-dotted line*: third approximation (sin(x) + 1/3 sin(3x) + 1/5 sin(5x)), *solid line*: fourth approximation (sin(x) + 1/3 sin (3x) + 1/5 sin(5x) + 1/7 sin(7x)). Note that not all harmonic frequencies are present and only sines are needed to represent the original antisymmetric signal. Box 3.1. explains how to determine the weights for each of the elements in the Fourier series.

Furthermore, if the signal is *symmetric* around time 0 only cosines (that are symmetric functions) are needed in the Fourier series, whereas when the signal is *antisymmetric* around zero (as in Fig. 3.11) only sines (that are antisymmetric functions) are needed to represent the signal. The more terms are added to the series, the more accurate the approximation of the original signal will be, as Fig. 3.11 also illustrates. In practice, only a few terms are often enough for a reasonable approximation of a periodic function/signal.

Calculating the coefficients (weights) for a Fourier series requires quite some mathematics but can be done (see Box 3.1 (Fourier Series) for mathematical details). By its weights, the Fourier series provides information on how important the different trigonometric functions are for representing the original function and as each trigonometric function has a specific frequency, the Fourier series provides information on the frequency content of the original function.

Box 3.1 Summary of the mathematics of Fourier series and Fourier transform (based on 'From Neurology to Methodology and back. An Introduction to Clinical Neuroengineering. Chapter Tremor, Polymyography, and Spectral Analysis, 2012, p. 43–44, Natasha Maurits. With permission of Springer.)

Fourier series
Any periodic function $x(t)$ of period T can be expressed as a Fourier series as follows:

$$x(t) = \sum_n a_n \sin(2\pi nt/T) + b_n \cos(2\pi nt/T), n = 0, 1, 2, \ldots \qquad (3.4)$$

Here, the frequency of each sine and cosine is n/T. Given $x(t)$, the coefficients a_n and b_n can be calculated by integrations over one period of $x(t)$:

(continued)

$$a_n = \frac{2}{T}\int_0^T x(t)\sin(2\pi nt/T)dt, n > 0, \quad a_0 = 0$$

$$b_n = \frac{2}{T}\int_0^T x(t)\cos(2\pi nt/T)dt, n > 0,$$

$$b_0 = \frac{1}{T}\int_0^T x(t)dt$$

Fourier transform

The Fourier transform is an extension of the Fourier series, but for non-periodic signals. A Fourier transform requires a functional description of the signal x(t) and also results in a functional (complex-valued) expression:

$$X(f) = \int_{-\infty}^{\infty} x(t)e^{-2\pi i f t}dt$$

Here, the sum in Eq. 3.4 has become an integral and the sines and cosines are now represented according to Euler's formula.

The inverse transform can also be executed and is referred to as the inverse Fourier transform:

$$x(t) = \int_{-\infty}^{\infty} X(f)e^{2\pi i f t}df$$

Unfortunately, real-world signals are hardly ever periodic, making it impossible to determine a Fourier series of such signals. These signals may be almost periodic over short periods of time, but variation over time is inevitable for signals that are obtained from real-world systems that are subject to noise of varying origin. Therefore, Fourier series do not provide a practical approach to get information on the frequency content of real-world signals. The Fourier transform (see Box 3.1 (Fourier transform) for its mathematical expression), can be used instead to assess the frequency content of non-periodic signals and is, in that sense, a generalization of the Fourier series. However, the Fourier transform can only be applied when the signal can be expressed as a function which is not the case for real-world signals that are normally sampled. Thus, we need another form of the Fourier transform to obtain the frequency content of sampled signals: the discrete Fourier transform (DFT). To perform the DFT efficiently using a computer, the transform has been implemented in several software programs as a fast Fourier transform (FFT), which makes Fourier analysis quite easy in practice. To be fast, most FFT implementations assume that the signal length N is a power of 2 (e.g. 512, 1024, 2048, ...). If that is the case the FFT algorithm will return exactly the same value(s) as a DFT. If that is not the case, the FFT algorithm will add values to the signal until its length has become the next power of two – a procedure referred to as zero padding.

To allow interpretation of the results of a DFT/FFT, it is necessary to understand some details of the transform. In a DFT of a signal of T seconds and consisting of N samples, for all

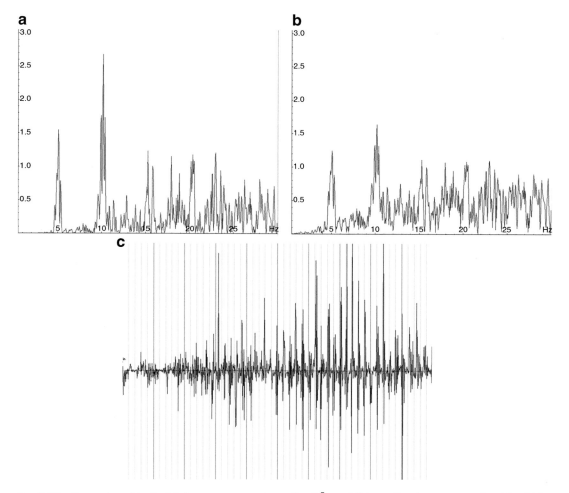

Fig. 3.12 Examples of both (**a**) the power spectrum (in μV²) and (**b**) amplitude spectrum (in μV) of (**c**) a piece of physiological signal (an *electroencephalogram*, measuring electrical brain activity), calculated by FFT. Notice that the peaks are relatively more pronounced in a power spectrum, which is therefore used more often in practice. However, to evaluate the relative contribution of different frequencies to the signal, an amplitude spectrum is more suitable. ('From Neurology to Methodology and back. An Introduction to Clinical Neuroengineering. Chapter Tremor, Polymyography, and Spectral Analysis, 2012, p. 46, Natasha Maurits. With permission of Springer.).

N sines and cosines with frequencies 0, $1/T$, $2/T$, $3/T$, 4 T, . . ., N-$1/T$ the extent to which they are present in the original signal is determined by a mathematical expression. Evaluation of this expression results in two coefficients or weights for each frequency: one for the sine and one for the cosine (quite similar to the Fourier series, see Box 3.1). Note that these coefficients can be zero, implying that the related frequency is not present in the signal. The result of this calculation is often displayed as a *spectrum* (see Fig. 3.12 for an example), with frequency on the horizontal axis. Some form of the coefficients for each frequency is set out vertically; for a power spectrum the sum of the squared coefficients is displayed, for an amplitude spectrum the square root of the values in the power spectrum is presented. The lower frequencies (longer periods) are closer to zero, to the left of the horizontal axis, while the higher frequencies (shorter periods) are to the right of the horizontal axis. Here, I would like to remind you of Eq. 3.2 that was used to explain that smaller values of B are related to smaller frequencies.

The frequencies in the spectrum are thus fully determined by the length T of the signal that is analyzed, as multiples of $1/T$. The highest frequency in the spectrum is the so-called *Nyquist frequency*, which is equal to half the sampling frequency. Or vice versa, the sampling frequency should be at least twice the highest frequency in the signal. This can intuitively be understood by realizing that we need at least two points to represent an oscillation, one at the maximum and one at the minimum. So, if the signal was sampled at 500 Hz (500 samples obtained every second), the maximum spectral frequency is 250 Hz. As the Fourier transform provides a complex-valued result, every trigonometric function contributing is not only characterized by its frequency (B in Eq. 3.2) but also by its phase (C in Eq. 3.2). Phase spectra can also be made (plotting the phase as a function of frequency), but when people talk about a spectrum, they usually refer to the magnitude (power or amplitude) spectrum.

3.4.1 An Alternative Explanation of Fourier Analysis: Epicycles

An intuitive way to derive the formula for the Fourier transform predates Fourier considerably, originating in ancient Alexandria in Egypt. Claudius Ptolemy, who lived around 150 CE, realized that the motion of celestial bodies was not perfectly circular as Aristotle's circular cosmology posed until then, but speeded up, slowed down, and in the cases of the planets even stopped and reversed (http://galileo.rice.edu/sci/theories/ptolemaic_system.html). One of Ptolemy's improved constructions to describe the motions includes an *epicycle* and *deferent* (Fig. 3.13). The motions described by epicycles and deferents look a bit like Spirograph drawings.

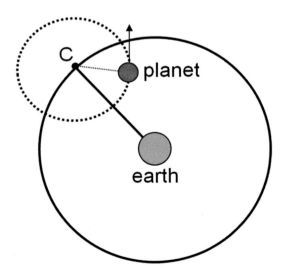

Fig. 3.13 From earth the planet in this diagram appears to be moving on a circle whose center moves around earth (Neugebauer, 1969). The *larger circle* is the deferent, the *smaller circle* the epicycle.

To further fine tune the description of the motions, this principle (putting circular trajectories on top of each other) can be repeated indefinitely. And actually, by doing this, any periodic movement can be represented. Funny examples can be viewed onYoutube (e.g. http://youtu.be/QVuU2YCwHjw). This should remind you of Fourier series, the infinite sum of trigonometric functions being able to represent any periodic function. So let's find out if there is indeed a relationship between epicycles, deferents and Fourier series.

Mathematically, a point that moves around on a circle with radius R can be described as if it were moving in the complex plane around the origin, very similar to the point moving on the unit circle in Fig. 3.7, albeit with non-unitary radius R. According to Euler's formula (Eq. 3.3) and Fig. 3.9 a point p that moves around a circle with radius R and angular frequency $\omega = 2\pi f$ is described by

$$p(t) = \mathrm{Re}^{i\omega t} = \mathrm{Re}^{2\pi i f t}$$

Now, if a point p moves around with angular frequency $\omega_1 = 2\pi f_1$ on a circle with radius R_1 that is itself moving around another circle with radius R_2 with angular frequency $\omega_2 = 2\pi f_2$, as in Fig. 3.13, this is described by:

$$p(t) = R_1 e^{2\pi i f_1 t} + R_2 e^{2\pi i f_2 t}$$

When extending this principle indefinitely, so that we get many circular trajectories on top of each other with any possible (angular) frequency, the sum turns into an integral (see also Chap. 7) and we arrive at:

$$p(t) = \int_{-\infty}^{\infty} R(f) e^{2\pi i f t} df$$

And this is precisely the definition of the inverse Fourier transform (cf. Box 3.1), where the trajectory in the time domain is described by $p(t)$ and in the frequency domain by $R(f)$. So here is the relationship between epicycles, deferents and Fourier series.

3.4.2 Examples and Practical Applications of Fourier Analysis

The simplest examples of FFTs of functions are those of trigonometric functions themselves: the FFT of a sine with frequency f in the time domain is given by a *delta function* at f Hz in the frequency domain (see Fig. 3.14a). As the Fourier transform is a linear transform, the FFT of the sum of two sines with frequencies f_1 and f_2 in the time domain is given by two delta functions at f_1 and f_2 Hz in the frequency domain (see Fig. 3.14b). In Fig. 3.11 an example was given of a Fourier series expansion of a block function; its FFT is given in Fig. 3.14c. What the examples in Fig. 3.14a, b and d illustrate is that finite *support* of the function in one domain (like a delta function) implies infinite support in the other domain

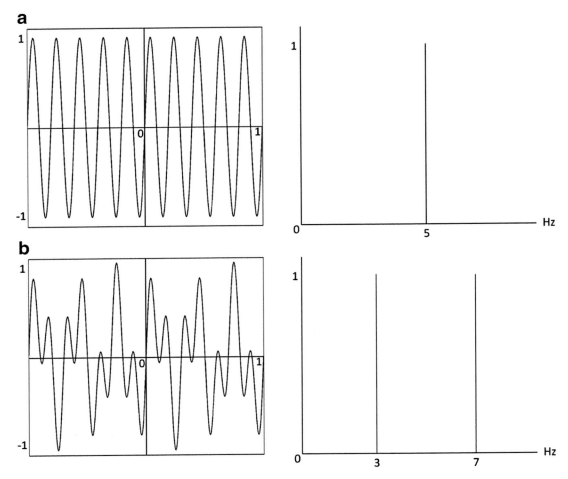

Fig. 3.14 Examples of Fourier transforms for simple functions. *Left*: function in the time domain, right: function in the frequency domain (**a**) Single sine with frequency 5 Hz. (**b**) Sum of two sines with frequencies 3 and 7 Hz. (**c**) Periodic block signal (Fig. 3.11). (**d**) (non-periodic) block signal with support $[-a,a]$. Its Fourier transform is given by $\frac{2\sin(af)}{f}$.

(like a sine). The reverse is not true; the (periodic) block function and its Fourier transform in Fig. 3.14c both have infinite support.

Exercise

3.13. Sketch the FFT of

   (a) a sine with frequency 3 Hz
   (b) the sum of two sines (with frequencies 2 Hz and 5 Hz)
   (c) white noise
   (d) 50 Hz noise

Fourier transforms also find many examples in the spectral analysis of biological signals; one example for an electroencephalogram (or EEG) was given in Fig. 3.14. The reason to

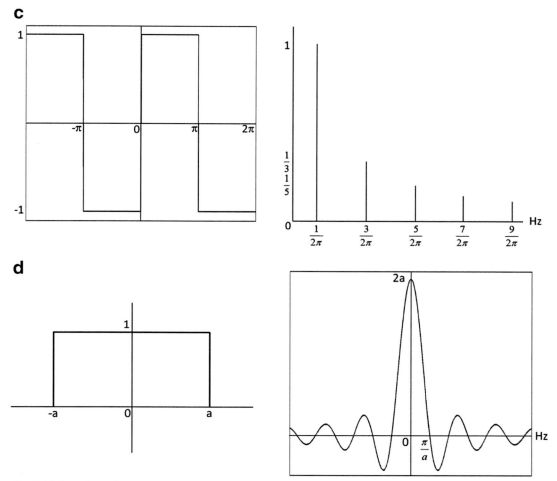

Fig. 3.14 (continued).

analyze biological signals in the frequency domain (as is done when calculating a spectrum) is often that there is relevant information to be obtained for clinical decision making. In the case of an EEG, the spectral power in the frequency band between 8 and 13 Hz (referred to as alpha band), for example, conveys information on how relaxed the patient is. The alpha peak frequency becomes smaller with age, indicating general slowing of brain activity.

Another example of obtaining clinically relevant information is the use of spectral analysis of *electromyography* (EMG) and *accelerometry* signals to assess *tremor*. By measuring the electrical activity of muscles (EMG) and the movement of the trembling limb itself, the peak tremor frequency and its variability across movements and postures can be obtained from the spectrum. This helps a clinician to make a distinction between different forms of tremor such as *essential tremor, enhanced physiological tremor* and *functional tremor*. Some key features to make this distinction are that the peak tremor frequency is typically more variable in functional tremor, and that there is often a shift in the peak tremor frequency for enhanced physiological tremor upon *loading* of the trembling limb. An example of accelerometry and electromyography spectra for a patient with functional tremor is given in Fig. 3.15, illustrating this variability in peak tremor frequency across different movements and postures.

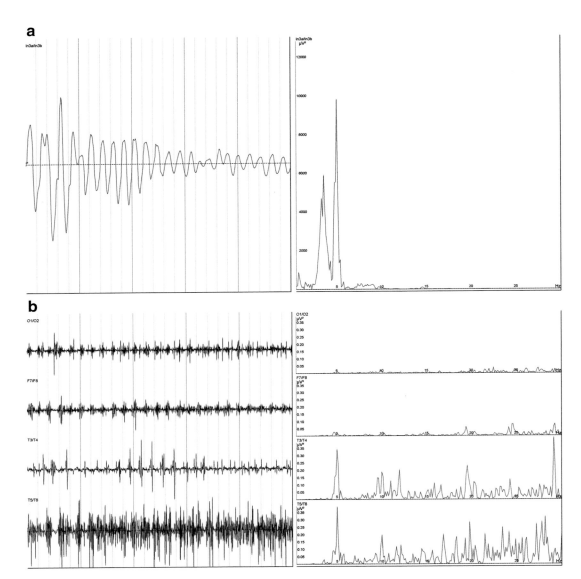

Fig. 3.15 Results of polymyography in a patient with functional tremor. *Left*: accelerometer or EMG traces. EMG from two muscle groups in the left arm and two muscle groups in the right arm (all traces displayed for 4–5 s), *right*: the Fourier spectrum for the selected data. Spectra are displayed from 0–30 Hz. Vertical axes have been adapted individually for each figure. (**a**) Accelerometer results during hand relaxation: tremor in right hand at 5.0 Hz. (**b**) EMG results (same time interval as **a**) during hand relaxation: tremor in the right hand at 5.0 Hz. Tremor bursts are visible in the EMG. (**c**) Accelerometer results during top-nose test: coarse tremor in right hand at 4.9 Hz. Slight tremor at the same frequency in the left hand. Note the triple *harmonic* peak in the spectrum at 14.7 Hz. (**d**) EMG results (same time interval as **c**) during top-nose test: tremor in the right hand at 4.9 Hz. Harmonics are visible in the spectra at two, three and multiple times this base frequency. (**e**) Accelerometer results during *diadochokinesis* with the left hand (*top signal*) at 1.2 Hz. A tremor develops in the right hand (*bottom signal*) at approximately the double frequency (2.6 Hz). (**f**) EMG results (same time interval as **e**) during diadochokinesis with the left hand at 1.2 Hz. The tremor that is observed in the accelerometer recording in (**e**) is not visible in the right hand EMG (spectrum) because of its irregularity. ('From Neurology to Methodology and back. An Introduction to Clinical Neuroengineering. Chapter Tremor, Polymyography, and Spectral Analysis, 2012, p. 53–55, Natasha Maurits. With permission of Springer.').

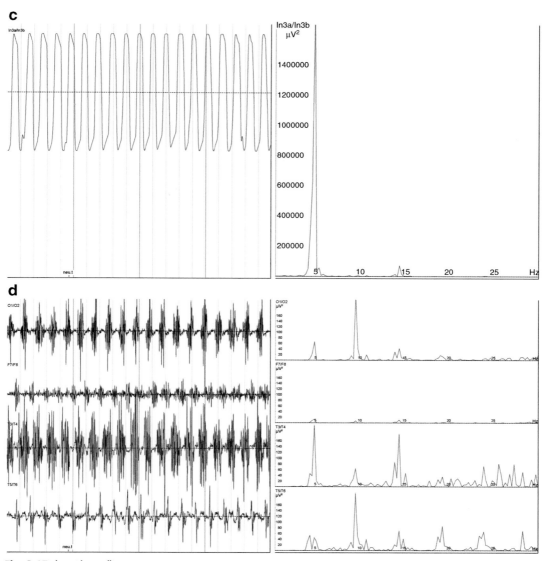

Fig. 3.15 (continued).

3.4.3 2D Fourier Analysis and Some of Its Applications

So far, I have considered one-dimensional Fourier transform only, for functions of one variable, such as time. However, Fourier transform can be easily extended to more dimensions, by applying the transform to one dimension at a time. The reason to talk about this here, is that the two-dimensional Fourier transform has some interesting and often-used applications in image analysis and neuroimaging.

I'll explain the 2D Fourier transform by applying it to the example grey-color image in Fig. 3.16a. To visualize and explain the result of the Fourier transform I will focus on the

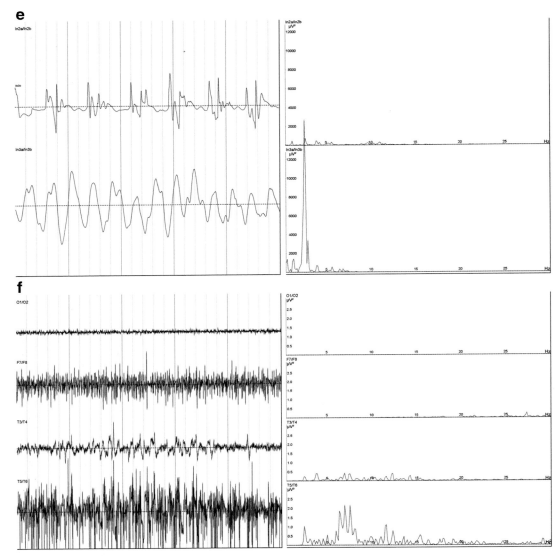

Fig. 3.15 (continued).

magnitude spectrum in Fig. 3.16b, which is comparable to the 1D spectral examples provided above.

In the magnitude spectrum, the zero frequency is now right in the middle, with the two spatial frequency axes cutting the magnitude spectrum in equal halves. In 2D Fourier transform, the contributing trigonometric functions are no longer 1D sines and cosines but 2D sinusoidal waves with different frequencies and orientations depending on the location in the magnitude spectrum (Fig. 3.17).

To get a better understanding of how the different spatial frequencies contribute to form the image, I first removed the lowest frequencies from the magnitude spectrum, setting their coefficients to zero (Fig. 3.18 top left). When inverse transforming the image, Fig. 3.18 top

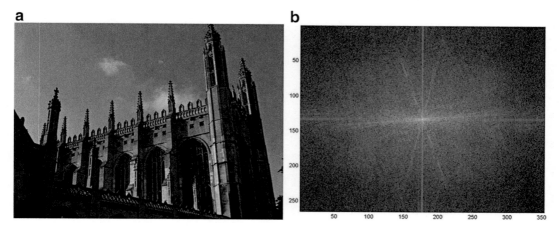

Fig. 3.16 (a) Example grey-color image and (b) the magnitude spectrum of its 2D Fourier transform.

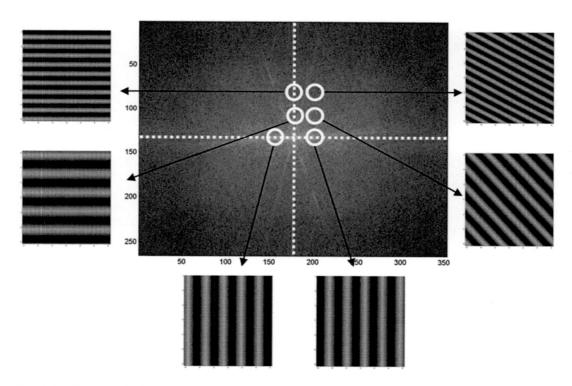

Fig. 3.17 The magnitude spectrum of Fig. 3.16b, with its two spatial frequency axes indicated and some representative contributing spatial sine waves. Locations on the axes correspond to vertically or horizontally oriented spatial sine waves, off-axes locations correspond to spatial sine waves at an angle. The closer to the origin, the smaller the spatial frequency and the larger the period.

Fig. 3.18 Effect of removing selected spatial frequencies from the magnitude spectrum in 3.16b of the original image in Fig. 3.16a. *Top left*: mask indicating which spatial frequencies were preserved (in white, Gaussian drop-off) for high-pass filtering. Note that in this representation the lowest frequencies are in the corners of the mask. *Top right*: Image with lowest spatial frequencies removed. Only fine details in the image remain. *Bottom left*: mask for low-pass filtering. *Bottom right*: Image with highest spatial frequencies removed. The image is blurred.

right results, illustrating that after removing the low spatial frequencies from the image, only the details of the image result. Vice versa, when removing the high spatial frequencies from the magnitude spectrum (Fig. 3.18 bottom left), after inverse Fourier transformation, only the coarse structures of the image remain; the image looks blurred (Fig. 3.18 bottom right).

The procedure described here is a form of spatial *filtering*: selected spatial frequencies are maintained in the image, while others are suppressed. More generally, data filtering (also in one dimension) can be performed efficiently by 'detouring' through the frequency domain (Fig. 3.19).

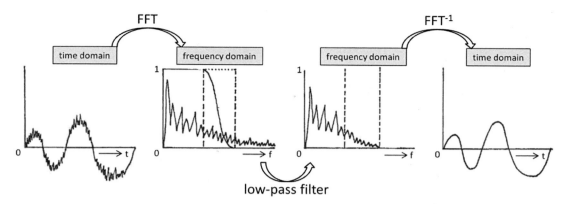

FFT

FFT⁻¹

Fig. 3.19 Principle of filtering via the frequency domain. In a first step the signal is transformed to the frequency domain using FFT. Then, in a second step, unwanted frequencies are suppressed by multiplication with the *transfer function* of a low-pass filter (to suppress high frequencies in this case) and finally, the signal is transformed back to the time domain using inverse FFT (FFT⁻¹). The transfer function is indicated in the second part of the figure; it is 1 for frequencies that need to be preserved, 0 for frequencies that need to be fully suppressed and smoothly varies from 1 to 0 in the transition band (between the dashed lines). *f*: frequency, *t*: time.

Instead of setting some frequencies to zero while maintaining others, there is usually a smooth transition between frequencies that are suppressed (multiplied by zero in the frequency domain) and frequencies that are maintained (multiplied by one in the frequency domain). This transition is reflected in a masking function which is known as the *transfer function*. Depending on which frequencies are suppressed and maintained, a filter can be low-pass (suppressing high frequencies; Fig. 3.20 top left), high-pass (suppressing low frequencies; Fig. 3.20 top right), band-pass (suppressing low and high frequencies; Fig. 3.20 bottom left) or band-stop or notch (suppressing a band of frequencies; Fig. 3.20 bottom right).

Filtering is one of the most common applications of the Fourier transform, that is applied in image processing (many options in graphical software like Photoshop or Paint employ spatial filters), but also in many other engineering and scientific fields as it provides a method for *data compression* and thus for efficient data storage. For example, the jpeg image format employs the (Fast) Fourier transform. But filters are also used in music to adapt its harmonic content and thereby change its timbre, or in cochlear implants to preprocess and denoise the incoming sound for optimal presentation to the patient.

A final example application of 2D Fourier transform is in neuroimaging: the collection of magnetic resonance imaging (MRI) data is fully done in Fourier space, or k-space, as it is also referred to. An MR image is only obtained after the k-space data is inverse Fourier transformed: this process is referred to as image reconstruction. To collect the raw data in k-space comes quite naturally to MR imaging of a body part as protons that are excited by radiofrequency waves and then subjected to magnetic gradient fields emit waves that can be read out. By cleverly applying magnetic gradients of different strengths and directions, a body part can be imaged slice by slice and every point in 2D k-space can be covered.

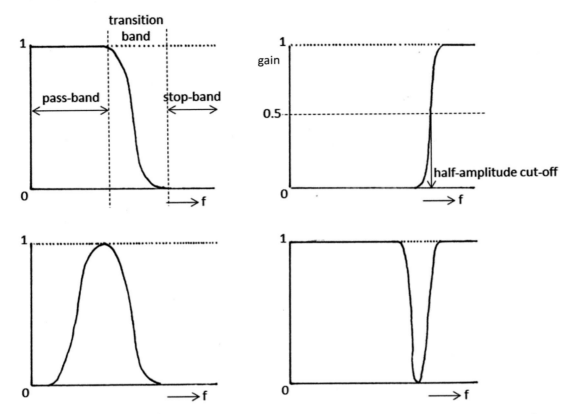

Fig. 3.20 Illustration of the four types of filters by their transfer functions. *Top left*: low-pass filter, top-right: high-pass filter, *bottom left*: band-pass filter, bottom-right: band-stop filter. *f*: frequency. The frequencies that are fully preserved are in the pass-band, while frequencies that are fully suppressed are in the stop-band. The area between the pass-band and the stop-band is referred to as the transition zone. A filter is often described by its half-amplitude cut-off (and by its roll-off; the steepness of the transfer function in the transition band), where the gain (the amplification factor of the transfer function) is 0.5. Hence, for example, for a 40 Hz low-pass filter, the half-amplitude cut-off is at 40 Hz.

Glossary

Accelerometry Measurement of acceleration using accelerometers (sensor device).

Antisymmetric In this chapter: a function f for which $f(-x) = -f(x)$.

Cosine (trigonometric) Ratio between the adjacent edge and the hypothenuse in a right-angled triangle.

Data compression To represent data such that it occupies less memory.

Deferent In specific movements described by epicycles and deferents the larger circle that an object moves along.

Delta function A function that is zero everywhere, except at zero, with an integral of 1 over the entire real domain.

Diadochokinesis Fast execution of opposite movements, e.g. rotating the hand from palm up to palm down and back.

Enhanced physiological tremor Tremor that occurs normally, e.g. when fatigued, but stronger in amplitude.

Electroencephalography Measurement of electrical brain activity using electrodes placed on the scalp.

Electromyography Measurement of electrical muscle activity using electrodes placed on the skin over a muscle (formally: surface electromyography).

Epicycle In specific movements described by epicycles and deferents the smaller circle that an object moves along.

Essential tremor Patients with this type of tremor typically present with action tremor (tremor during posture and/or movements), mostly in the arms, with tremor frequency in the range of 4–12 Hz.

Filtering Suppression of specific frequencies in a signal.

Function A mathematical relation, like a recipe, describing how to get from an input to an output.

Functional tremor Tremor for which no organic cause can be found.

Hypothenuse Edge opposite the right angle in a right-angled triangle.

Modulo Modulo(n) expresses equivalence up to a remainder n.

Nyquist frequency The highest frequency that is still adequately represented in a sampled signal (half the sampling frequency).

Periodic Repeating itself regularly.

Right-angled (triangle) A geometric object (triangle) with a right (90°) angle.

Similar (triangles) Triangles that have the same angles but differ in size.

SI Système international d'unités (international system of units).

Sine (trigonometric) Ratio between the opposite edge and the hypothenuse in a right-angled triangle.

Spectrum Here: representation of a function or signal as a function of frequency, typically resulting from a Fourier transform.

Support The support of a function is the set of points where the function values are non-zero.

Symmetric In this chapter: a function f for which $f(-x) = f(x)$.

Tangent (trigonometric) Ratio between the opposite edge and the adjacent edge in a right-angled triangle.

Transfer function Here: expresses how strongly frequencies are preserved or suppressed by a filter, can vary between 0 and 1.

Tremor Oscillatory movement of a body part.

Unit circle Circle with radius one.

Symbols Used in This Chapter (in Order of Their Appearance)

sin	Sine
cos	Cosine
tan	Tangent
$\lim_{x \to 0}$	Limit for x approaching zero
α, θ, ϕ	Angles
r, R	Radius (of a circle)
rad	Radian
e	Exponential
i	Imaginary unit
$\sum$	Sum
$\int$	Integral
f	Frequency
t	Time
ω	Angular frequency

Overview of Equations, Rules and Theorems for Easy Reference

Mnemonic for memorizing trigonometric ratios

sine   = O/H
cosine   = A/H
tangent   = O/A = sine/cosine
where O = opposite, A = adjacent, H = hypothenuse

Angles and values to remember

x in rad	$\pi/6$	$\pi/4$	$\pi/3$	$\pi/2$	π
$\sin(x)$	0.5	$\frac{1}{2}\sqrt{2}$	$\frac{1}{2}\sqrt{3}$	1	0

Euler's formula

$$e^{ix} = \cos(x) + i\sin(x)$$

Trigonometric relations

$$\sin^2(x) + \cos^2(x) = 1$$

$$\cos(x+y) = \cos(x)\cos(y) - \sin(x)\sin(y)$$
$$\sin(x+y) = \cos(x)\sin(y) + \sin(x)\cos(y)$$

$$\cos(x-y) = \cos(x)\cos(y) + \sin(x)\sin(y)$$
$$\sin(x-y) = \sin(x)\cos(y) - \cos(x)\sin(y)$$

$$\sin(2x) = 2\sin(x)\cos(x)$$
$$\cos(2x) = \cos^2(x) - \sin^2(x)$$

$$\cos(2x) = 1 - 2\sin^2(x)$$

Answers to Exercises

3.1. The ratios must be calculated for the angle θ, so the sine, e.g., is always the length of the edge opposite from θ divided by the length of the hypothenuse etcetera. Sometimes rotating the triangles in your mind to relate them to the triangles in Fig. 3.3 will help. Thus the answers are for the left triangle: sine = 4/5, cosine = 3/5, tangent = 4/3, middle triangle: sine = 4/7, cosine = $\sqrt{33}/7$, tangent = $4/\sqrt{33}$, right triangle: sine = 3/5, cosine = 4/5, tangent = 3/4.

3.2 First the lengths of the edges of the two triangles that angles 1 to 4 belong to must be determined. For angle 1, the adjacent edge is 75 cm long; this is the opposite edge for angle 2. For angle 2, the adjacent edges is 45 cm long; this is the opposite edge for angle

1. Similarly, the adjacent edge for angle 3 (opposite edge angle 4) is 65 cm long and the adjacent edge for angle 4 (opposite edge angle 3) is 35 cm long. The angles can now be determined by using the trigonometric ratio for the tangent: tan = O/A. Thus, $\tan \angle 1 = \frac{45}{75}$, $\tan \angle 2 = \frac{75}{45}$, $\tan \angle 3 = \frac{35}{65}$ and $\tan \angle 4 = \frac{65}{35}$. A calculator can now be used to determine the angles using the inverse tangent (or atan) function.

3.3 (a) $30° = 30*2\pi/360 = \pi/6$ rad
    (b) $45° = 45*2\pi/360 = \pi/4$ rad
    (c) $60° = 60*2\pi/360 = \pi/3$ rad
    (d) $80° = 80*2\pi/360 = 4/9 \, \pi$ rad
    (e) $90° = 80*2\pi/360 = \pi/2$ rad
    (f) $123° = 123*2\pi/360 = 123/180 \, \pi$ rad
    (g) (g) $260° = 260*2\pi/360 = 13/9 \, \pi$ rad
    (h) (h) $-16° = -16*2\pi/360 = -4/45 \, \pi$ rad
    (i) $-738° = -738 + 2*360° = -18° = -18*2\pi/360 = -1/10 \, \pi$ rad

3.4. (a) $2\pi/3$ rad $= 2\pi/3*360/2\pi° = 120°$
    (b) $\pi/4$ rad $= \pi/4*360/2\pi° = 45°$
    (c) $9\pi/4$ rad $= 9\pi/4-2\pi$ rad $= \pi/4*360/2\pi° = 45°$
    (d) 0.763π rad $= 0.763 \, \pi *360/2\pi° = 137{,}34°$
    (e) π rad $= \pi*360/2\pi° = 180°$
    (f) $\theta\pi$ rad $= \theta\pi*360/2\pi° = 180\theta°$
    (g) θ rad $= \theta*360/2\pi° = 180\theta/\pi°$

3.5. (a) false: $\cos 0° = \cos (0 \text{ rad}) = 1$
    (b) false: $\sin 30° = \sin (\pi/6 \text{ rad}) = 0.5$
    (c) false: $\sin 45° = \sin (\pi/4 \text{ rad}) = \frac{1}{2}\sqrt{2}$
    (d) true
    (e) false: $\cos 60° = \cos(\pi/3 \text{ rad}) = 0.5$

3.6. For the angle of 60° the length of the opposite edge and the hypothenuse are x and 13, respectively. These two values can thus be used to calculate the sine of this angle by their ratio. But the sine of this angle is also equal to $\sin(60°) = \sin(\pi/3 \text{ rad}) = \frac{1}{2}\sqrt{3}$. Thus $x/13 = \frac{1}{2}\sqrt{3}$, or $x = \frac{13}{2}\sqrt{3}$.

3.7. See Fig. 3.8 for the graphs of $\sin(x)$, $\cos(x)$ and $\tan(x)$. The exact values for the angles 0, 30, 45, 60 and 90 degrees (that help you make these sketches) are:

degrees	rad	sin	cos	tan
0	0	0	1	0
30	$\pi/6$	0.5	$\frac{1}{2}\sqrt{3}$	$\frac{1}{3}\sqrt{3}$
45	$\pi/4$	$\frac{1}{2}\sqrt{2}$	$\frac{1}{2}\sqrt{2}$	1
60	$\pi/3$	$\frac{1}{2}\sqrt{3}$	0.5	$\sqrt{3}$
90	$\pi/2$	1	0	∞

3.8. (a) $\cos(-2x) = \cos(2x)$
 (b) $\tan(-\pi/4) = -\tan(\pi/4)$
 (c) $\sin(-4\pi/3) = -\sin(4\pi/3)$

3.9. $\sin(x)$: black, $\sin(2x)$: green, $\sin(x)+2$: red, $\sin(x+2)$: magenta and $2\sin(x)$: blue.

3.10. (a) For sketching $y = 4\sin(x)$, first make a table, with easily obtained values (making use of Table 3.1 and the trigonometric symmetries), e.g.:

x	y
$-\pi \cong -3.14$	0
$-\pi/2$	-4
$-\pi/3$	$-2\sqrt{3} \cong -3.46$
0	0
$\pi/3$	$2\sqrt{3} \cong 3.46$
$\pi/2$	4
π	0

Then sketch an x- and a y-axis to scale that are long enough to accommodate for the maximum (4) and minimum (-4) of the function and for at least one period ($2\pi \cong 6.28$) of the function. Note that you are free to choose the scale of the axes. Put in the points you calculated in the table and sketch a smooth line through the points. Such a sketch could look like:

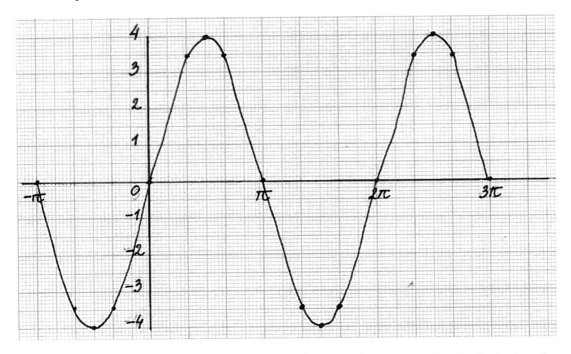

Note that a sketch of a trigonometric function does not need to be limited to the positive x-axis: I here sketched it for both negative and positive values of x and also for more than one period.

(b) For sketching $y = \sin(4x)$, first make a table, with easily obtained values (making use of Table 3.1 and the trigonometric symmetries), e.g.:

x	y
$-\pi/4$	0
$-\pi/8$	-1
$-\pi/12$	$-\frac{1}{2}\sqrt{3} \cong -0.87$
0	0
$\pi/12$	$\frac{1}{2}\sqrt{3} \cong 0.87$
$\pi/8$	1
$\pi/4$	0

Then sketch an x- and a y-axis to scale that are long enough to accommodate for the maximum (1) and minimum (-1) of the function and for at least one period ($\pi/2 \cong 1.57$) of the function. Put in the points you calculated in the table and sketch a smooth line through the points. Such a sketch could look like:

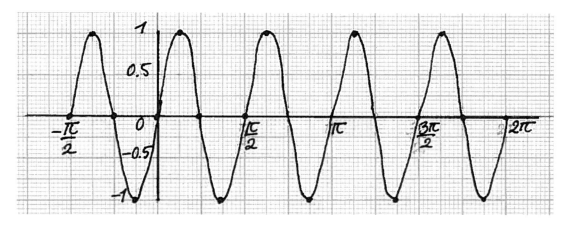

3.11. (a) Following a similar procedure as in Exercise 3.10 a sketch of $y = 2 + \sin(x)$ could look like:

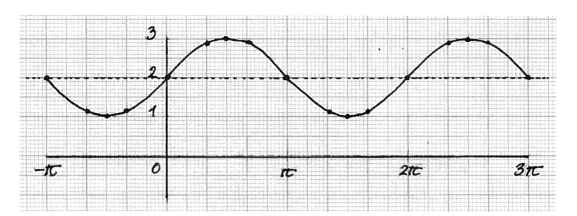

(b) Similarly, a sketch of $y = -\cos(x)$ could look like:

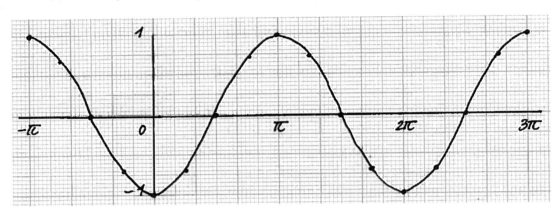

3.12. (a) $y = -3\sin(x) - 5$ is obtained from the standard sine function by multiplying it by -3 and shifting it down by 5.

(b) $y = 2\sin(3x)$ is obtained by multiplying it by 2 and changing the period to $2\pi/3$ (if this last part is hard to grasp: the period is that number that you fill in for x, making the total between the brackets equal to 2π).

(c) $y = 2\sin(x + \pi) + 2$ is obtained by multiplying it by 2, shifting it up by 2 and shifting it to the left by π (again, it is to the left and not to the right because when I fill in $-\pi$, the value between brackets becomes 0).

(d) $y = 3\sin\left(2x - \frac{\pi}{2}\right) - 1$ is obtained by multiplying it by 3, shifting it down by 1, changing the period to π and shifting it to the right by $\pi/2$.

(e) $y = -4\sin\left(\frac{\pi x}{5}\right)$ is obtained by multiplying it by -4 and changing the period to 10.

(f) $y = -x + 3\sin(2x - \pi)$ is obtained by multiplying it by 3, changing the period to π, shifting it to the right by π and finally by plotting this whole function around the line $y = -x$. It will look like:

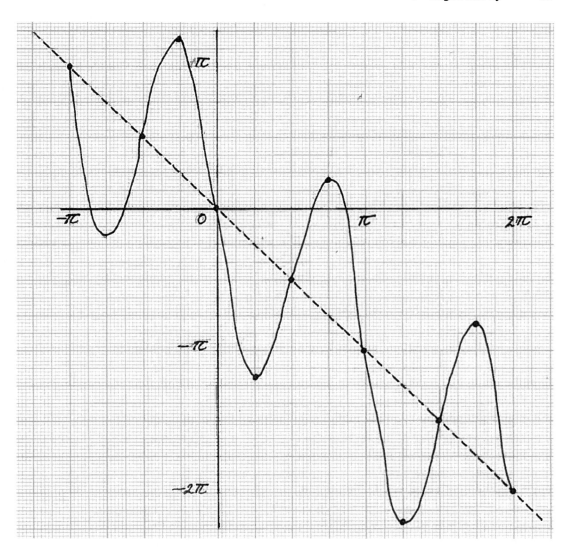

3.13. The FFT of

(a) a sine with frequency 3 Hz looks like:

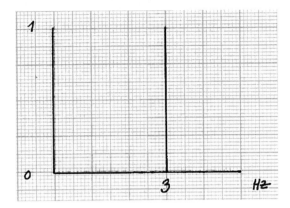

(b) the sum of two sines (2 Hz and 5 Hz) looks like:

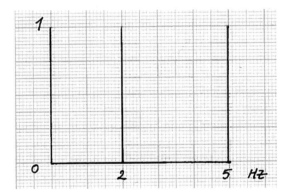

(c) white noise looks like (random spectrum):

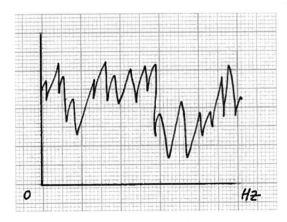

(d) 50 Hz noise looks like:

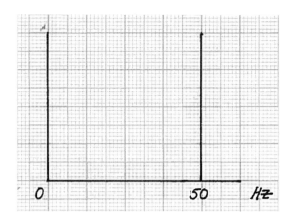

References

Online Sources of Information: Methods

https://en.wikipedia.org/wiki/Trigonometry
http://acronyms.thefreedictionary.com/Sohcahtoa
https://en.wikipedia.org/wiki/Fourier_analysis
http://galileo.rice.edu/sci/theories/ptolemaic_system.html (Epicycles)
Unbaking a cake by Carl Joseph (Alternative non-mathematical Fourier series explanation) https://www.youtube.com/watch?v=Qm84XIoTy0s&feature=youtu.be
http://math.stackexchange.com/questions/1002/fourier-transform-for-dummies (thread on relation between Epicycles and Fourier analysis)

Books

Otto Neugebauer – The exact sciences in antiquity. Courier Corporation, 1969. Available from books.google.com.

Papers

F.A. Farris, Wheels on wheels on wheels – surprising symmetry. Math. Mag. **69**(3), 185–189 (1996)
D.B. Plewes, W. Kucharczyk, Physics of MRI: a primer. J. Magn. Res. Imaging **35**, 1038–1054 (2012)

4

Vectors

Natasha Maurits

After reading this chapter you know:

- what vectors are and how they can be used,
- how to algebraically perform and geometrically represent addition and subtraction of vectors,
- how to algebraically perform and intuitively understand common forms of vector multiplication and
- how vector relationships can be expressed and what their use is.

4.1 What Are Vectors and How Are They Used?

Vectors, in mathematics, are entities that have both direction and magnitude, but not a fixed position. In two- and three-dimensional space vectors can be thought of and represented as arrows. Examples of vectors are forces and velocities. Interestingly, mathematical operations on vectors already existed before vectors were called vectors; the *parallelogram* law of adding vectors was already known to ancient scholars who studied physical problems. For example, the net force resulting from two forces that act on an object in different directions can be obtained as the diagonal of the parallelogram spanned by the two vectors (Fig. 4.1).

The name 'vector' was introduced only relatively recently, in the nineteenth and twentieth centuries, even though people had been working with *coordinates*, pairs of ordered numbers that have a close relationship with vectors, for much longer. Let me explain this relationship. Any point P in space is fixed by its coordinates, in case of 2D space by the pair (x,y) and in

N. Maurits (✉)
Department of Neurology, University Medical Center Groningen, Groningen, The Netherlands
e-mail: n.m.maurits@umcg.nl

© Springer International Publishing AG 2017
N. Maurits, B. Ćurčić-Blake, *Math for Scientists*, DOI 10.1007/978-3-319-57354-0_4

case of 3D space by the triplet (x,y,z). Here, x, y, and z are the coordinates of P. In vector notation this is represented as:

$$\vec{p} = \begin{pmatrix} x \\ y \end{pmatrix} \text{ in 2D space or } \vec{p} = \begin{pmatrix} x \\ y \\ z \end{pmatrix} \text{ in 3D space; here } x, y \text{ and } z \text{ are the } \textit{elements} \text{ of } \vec{p}.$$

In case of the latter notation, the vector can be represented as an arrow; a line piece with a direction that starts in the origin and ends at point P (Fig. 4.2). This physical arrow is still visible in the notation for vectors that we use here, where we draw a small arrow over the letter.

In two or three dimensions, vectors are sometimes also denoted in the 'ij' or 'ijk' notation, i.e. the vector $\vec{p}$ above would be represented as $\vec{p} = x\vec{i} + y\vec{j}$ in 2D or $\vec{p} = x\vec{i} + y\vec{j} + z\vec{k}$ in 3D. Here, $\vec{i}$, $\vec{j}$ and $\vec{k}$ are vectors of length 1 along the principal x-, y-, and z-axes, respectively.

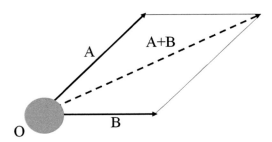

Fig. 4.1 The parallelogram law for vector addition illustrated for the case of a net force acting on an object O as the result of two forces A and B. The net force is the *dashed vector* A + B.

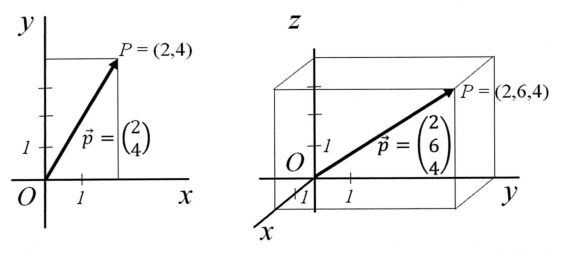

Fig. 4.2 Illustration of the relationship between a point P in 2D space (*left*) and in 3D space (*right*) and its associated vector $\vec{p}$.

4.2   Vector Operations

All common mathematical operations, such as addition, subtraction and multiplication, can be executed on vectors, just as they can be executed on e.g. real numbers. The extension of these basic operations to vectors is done in such a way that important properties of the basic mathematical operations (*commutativity*, *associativity* and *distributivity*, see Box 4.1) are preserved as much as possible.

Box 4.1 Properties of binary mathematical operations (examples)

Commutativity: $a + b = b + a$ $ab = ba$

Associativity: $a + (b + c) = (a + b) + c$ $a(bc) = (ab)c$

Distributivity: $a(b + c) = ab + ac$, when a is a *scalar*

In this section, the basic vector operations are explained both *algebraically* and *geometrically*.

4.2.1   Vector Addition, Subtraction and Scalar Multiplication

Let's start simply, with vector addition. This vector operation was geometrically already introduced in Fig. 4.1: addition of two vectors (in 2D or 3D space) is equal to taking the diagonal of the parallelogram spanned by the two vectors. Note that also in 3D space the parallelogram spanned by two vectors forms a plane. Adding three vectors can also be explained geometrically, as this is equal to taking the diagonal of the *parallelepipid* spanned by the three vectors (Fig. 4.3).

In more dimensions, or when adding more than three vectors, however, the geometrical structures needed for vector addition, become difficult to imagine. Alternatively, vectors can (algebraically) be added (or subtracted) by adding (or subtracting) their elements.

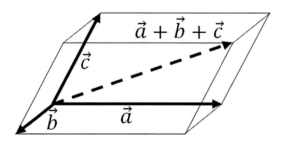

Fig. 4.3 Geometric explanation of adding three vectors $\vec{a}$, $\vec{b}$ and $\vec{c}$ in 3D space. The sum is the diagonal of the parallelepiped spanned by the three vectors.

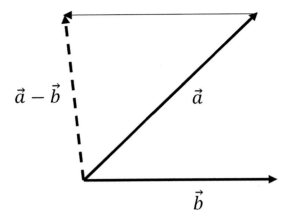

Fig. 4.4 Geometric explanation of subtracting two vectors $\vec{a}$ and $\vec{b}$ in 2D space. Their difference is the diagonal of the parallelogram spanned by the first vector and the negative of the second vector (indicated by a *thin line*), i.e. $\vec{a} - \vec{b} = \vec{a} + (-\vec{b})$.

For example, given the vectors $\vec{a} = \begin{pmatrix} a_1 \\ a_2 \end{pmatrix}$ and $\vec{b} = \begin{pmatrix} b_1 \\ b_2 \end{pmatrix}$, their sum $\vec{a} + \vec{b}$ equals $\begin{pmatrix} a_1 + b_1 \\ a_2 + b_2 \end{pmatrix}$ and similarly, their difference $\vec{a} - \vec{b}$ equals $\begin{pmatrix} a_1 - b_1 \\ a_2 - b_2 \end{pmatrix}$. This also holds true in more than two dimensions and for more than two vectors, e.g., in n-dimensional space:

$$\vec{a} + \vec{b} + \vec{c} = \begin{pmatrix} a_1 \\ \vdots \\ a_n \end{pmatrix} + \begin{pmatrix} b_1 \\ \vdots \\ b_n \end{pmatrix} + \begin{pmatrix} c_1 \\ \vdots \\ c_n \end{pmatrix} = \begin{pmatrix} a_1 + b_1 + c_1 \\ \vdots \\ a_n + b_n + c_n \end{pmatrix}.$$

Geometrically, vector subtraction is achieved by taking the diagonal of the parallelogram spanned by the first vector and the negative of the second vector, i.e. the vector of the same length, but with the opposite direction (Fig. 4.4).

Thus, algebraic and geometric addition and subtraction of vectors give the same results, albeit in a different notation: the algebraic notation uses arrays of numbers while the geometric notation employs arrows in a plane or in 3D space. Importantly, however, the algebraic notation can be generalized to more dimensions, where we cannot draw arrows anymore and the geometric approach fails.

Example 4.1

Add and subtract (algebraically) the pair of vectors $\begin{pmatrix} 7 \\ 3 \\ -2 \end{pmatrix}$ and $\begin{pmatrix} 2 \\ -5 \\ 8 \end{pmatrix}$.

(continued)

Example 4.1 (continued)

To add the vectors the elements of the two vectors are added: $\begin{pmatrix} 7 \\ 3 \\ -2 \end{pmatrix} + \begin{pmatrix} 2 \\ -5 \\ 8 \end{pmatrix} =$ $\begin{pmatrix} 7+2 \\ 3-5 \\ -2+8 \end{pmatrix} = \begin{pmatrix} 9 \\ -2 \\ 6 \end{pmatrix}$. To subtract the vectors the elements are subtracted: $\begin{pmatrix} 7 \\ 3 \\ -2 \end{pmatrix} - \begin{pmatrix} 2 \\ -5 \\ 8 \end{pmatrix} = \begin{pmatrix} 7-2 \\ 3+5 \\ -2-8 \end{pmatrix} = \begin{pmatrix} 5 \\ 8 \\ -10 \end{pmatrix}$.

From the definition of vector subtraction it follows that there must be a *null vector*, the result of subtracting a vector from itself. This is the only vector that has zero magnitude, which is why it is called the null vector.

Another basic vector operation is scalar multiplication. Multiplying an n-dimensional vector by a scalar (real number) s is defined as:

$$s\vec{a} = s\begin{pmatrix} a_1 \\ \vdots \\ a_n \end{pmatrix} = \begin{pmatrix} sa_1 \\ \vdots \\ sa_n \end{pmatrix}.$$

This can also be explained geometrically as illustrated in Fig. 4.5.

Finally, the representation of a vector as a point in space immediately suggests how we can define its magnitude (also referred to as length, module or absolute value), using Pythagoras theorem:

$$|\vec{a}| = \left| \begin{pmatrix} a_1 \\ \vdots \\ a_n \end{pmatrix} \right| = \sqrt{a_1^2 + \cdots + a_n^2}.$$

This is illustrated for 2D space in Fig. 4.6.

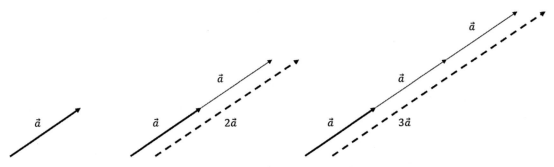

Fig. 4.5 Geometric explanation of scalar multiplication of a vector $\vec{a}$ (*left*) by a factor 2 (*middle*) and 3 (*right*). The result is indicated by *dashed vectors*. Thus scalar multiplication is equal to multiple additions.

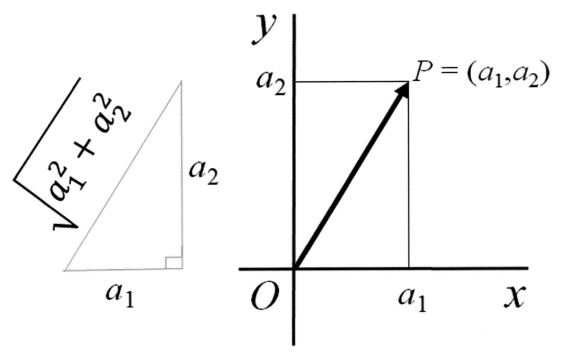

Fig. 4.6 Geometric explanation of vector magnitude in 2D space. The *left triangle* is equal to the triangle formed by *O-a₁-P* in the *right part* of the figure.

Exercises

4.1. Add and subtract (i.e. first-second vector) the following pairs of vectors:

(a) $\begin{pmatrix} 7 \\ 3 \end{pmatrix}$ and $\begin{pmatrix} 2 \\ 4 \end{pmatrix}$

(b) $\begin{pmatrix} -2 \\ 17 \end{pmatrix}$ and $\begin{pmatrix} 3 \\ -14 \end{pmatrix}$

(c) $\begin{pmatrix} 1 \\ 2 \\ 3 \end{pmatrix}$ and $\begin{pmatrix} 4 \\ -23 \\ 7 \end{pmatrix}$

(d) $\begin{pmatrix} 2.4 \\ 1.2 \\ 3.6 \\ 5.4 \end{pmatrix}$ and $\begin{pmatrix} 1.2 \\ 3.6 \\ 5.4 \\ 2.4 \end{pmatrix}$

4.2. For Exercise 4.1(a), also perform the addition and subtraction geometrically.

4.3. Calculate

(a) $\begin{pmatrix} 1 \\ 3 \\ 2 \end{pmatrix} + 3\begin{pmatrix} -3 \\ 2 \\ -5 \end{pmatrix} - 2\begin{pmatrix} -3 \\ 3 \\ 4 \end{pmatrix}$

(continued)

(b) $-\begin{pmatrix} 0.5 \\ 3.1 \\ 6.7 \end{pmatrix} - 2\begin{pmatrix} 1 \\ 0.7 \\ 1.2 \end{pmatrix} + 1.5\begin{pmatrix} 4 \\ 3 \\ 8 \end{pmatrix}$

4.4. Let $P = (-2,-2)$, $Q = (-3,4)$ and $R = (1,2)$ be three points in the xy-plane.

(a) Draw the points P, Q and R in the xy-plane and the vectors $\vec{u}$ joining P to Q, $\vec{v}$ joining Q to R and $\vec{w}$ joining R to P.
(b) What are the elements of the vectors of $\vec{u}$, $\vec{v}$ and $\vec{w}$?
(c) What is $\vec{u} + \vec{v} + \vec{w}$? First answer without calculation and verify by calculation.

4.5. An airplane is taking off in a straight line with velocity vector $\begin{pmatrix} 200 \\ 180 \\ 100 \end{pmatrix}$ (in km/h). Suppose that (x, y) are its ground coordinates and z its altitude.

(a) Where is the airplane after 1 min?
(b) How many seconds does it take the airplane to climb 100 m?
(c) How long would it take the airplane to reach its cruising altitude of 9 km if it would keep flying at this speed and in this direction?

4.2.2 Vector Multiplication

Let's take vector operations to the next level by considering multiplication of two vectors. This operation is a little less straightforward than the operations explained in the previous section, because there are multiple ways in which two vectors can be multiplied and they all have their specific applications. Here, only the often used inner or dot product and the cross product are discussed.

4.2.2.1 Inner Product

Geometrically, the inner product of two vectors $\vec{a}$ and $\vec{b}$ is defined as:

$$\vec{a} \cdot \vec{b} = |\vec{a}||\vec{b}|\cos\varphi,$$

where $0 < \varphi < \pi$ is the angle between the two vectors when they originate at the same position (Fig. 4.7).

This definition immediately shows that the inner product of two vectors is a scalar and that it is zero when the cosine of the angle between them is zero, i.e. when they are perpendicular. The reverse is also true. Hence, two vectors are perpendicular if and only if (*iff*) their inner product is zero. Another word for perpendicular is *orthogonal*. It is not very difficult to gain an intuitive understanding of the inner product. Remember that the cosine can be calculated as the ratio between the adjacent edge and the hypothenuse in a right-angled triangle (Sect. 3.2). Hence, in the triangle drawn in Fig. 4.8, $\cos\varphi$ equals the part of $\vec{a}$ in the direction of $\vec{b}$ (the adjacent edge *adj*), divided by $|\vec{a}|$.

This makes the inner product equal to:

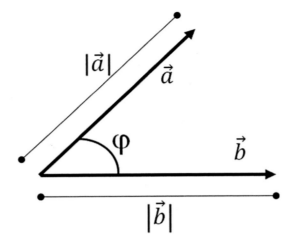

Fig. 4.7 Illustration of the geometric definition of the inner product of two vectors $\vec{a}$ and $\vec{b}$.

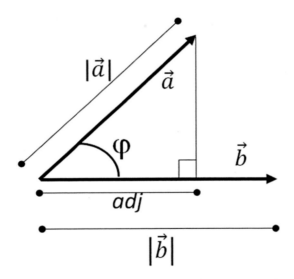

Fig. 4.8 Illustration supporting the intuitive understanding of the inner or dot product of two vectors $\vec{a}$ and $\vec{b}$, adj = adjacent edge.

$$\vec{a} \cdot \vec{b} = |\vec{a}||\vec{b}| \cos \varphi = |\vec{a}||\vec{b}|\frac{adj}{|\vec{a}|} = |\vec{b}|adj, \text{ hence to the part of } \vec{a} \text{ in the direction of } \vec{b},$$

times the length of $\vec{b}$. Thus, the more two vectors $\vec{a}$ and $\vec{b}$ are oriented in the same direction, the larger their inner product will be, with the maximum possible inner product being $|\vec{a}||\vec{b}|$.

This fits with the inner product being zero when two vectors $\vec{a}$ and $\vec{b}$ are orthogonal: in that case the part of $\vec{a}$ in the direction of $\vec{b}$ is zero. An application that illustrates this intuitive understanding is how *work* is calculated in physics. Work equals the product of the force exerted on an object and its displacement. If the (constant) force acts in the same direction as the displacement of the object (such as when pushing a spring), then this product is just a 'simple' product. If the force is not in the same direction as the displacement (e.g. in the case of friction), the work is the inner product of the force and the displacement, i.e. the product of the component of the force in the direction of the displacement and the displacement.

In higher dimensions, the geometrical definition of the inner product is again difficult to apply and luckily, there is also a simple algebraic definition:

$$\vec{a} \cdot \vec{b} = \begin{pmatrix} a_1 \\ \vdots \\ a_n \end{pmatrix} \cdot \begin{pmatrix} b_1 \\ \vdots \\ b_n \end{pmatrix} = a_1 b_1 + \cdots a_n b_n.$$

This definition implies that the magnitude (length, absolute value) of a vector can also be expressed in terms of the inner product with itself (cf. Sect. 4.2.1):

$$|\vec{a}| = \sqrt{a_1^2 + \cdots + a_n^2} = \sqrt{\vec{a} \cdot \vec{a}}$$

This definition of the length of a vector is also known as the Euclidean length, Euclidean *norm* or L_2-norm of a vector. When dividing a vector by its own length a *unit vector* results, i.e. with length 1. Thus, $\vec{i}$, $\vec{j}$ and $\vec{k}$, as introduced in Sect. 4.1 are unit vectors.

Another application of the inner product is how a plane is defined. Any plane in 3D through the origin is defined by $\vec{n} \cdot \vec{x} = 0$, where $\vec{n} = n_1 \vec{i} + n_2 \vec{j} + n_3 \vec{k}$ is a *normal vector*, perpendicular to this plane and $\vec{x}$ is any 3D vector such that $\vec{n} \cdot \vec{x} = 0$ (Fig. 4.9). Any plane parallel to this plane is also defined by this same normal vector <u>and</u> by one point in the plane that determines the (non-zero) constant on the right hand side of the equation. This is explained by an exercise in this section. The general definition of a plane as $\vec{n} \cdot \vec{x} = C$, where C is a constant, holds true for any dimension.

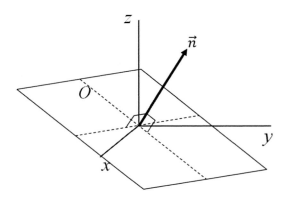

Fig. 4.9 A plane in 3D through the origin and its normal vector. Note that the normal vector is perpendicular to the plane and thus to any line in the plane.

Exercises

4.6. Calculate the inner product of these pairs of vectors algebraically

(a) $\begin{pmatrix} 1 \\ 0 \end{pmatrix}$ and $\begin{pmatrix} 0 \\ 1 \end{pmatrix}$

(b) $\begin{pmatrix} 1 \\ 1 \end{pmatrix}$ and $\begin{pmatrix} 1 \\ -1 \end{pmatrix}$

(c) $\begin{pmatrix} 1 \\ 2 \end{pmatrix}$ and $\begin{pmatrix} 3 \\ 6 \end{pmatrix}$

(d) $\begin{pmatrix} 2 \\ 1 \end{pmatrix}$ and $\begin{pmatrix} 1 \\ 2 \end{pmatrix}$

(e) $\begin{pmatrix} 2 \\ 1 \end{pmatrix}$ and $\begin{pmatrix} -1 \\ 3 \end{pmatrix}$

Note how the algebraic manner of calculating the inner product makes matters so much easier (compared to the geometrical calculation according to Fig. 4.8), especially in higher dimensions of course, as in the next exercise.

4.7. Calculate the inner product of these pairs of vectors algebraically

(a) $\begin{pmatrix} 1 \\ 2 \\ 3 \end{pmatrix}$ and $\begin{pmatrix} 4 \\ 5 \\ 7 \end{pmatrix}$

(b) $\begin{pmatrix} 1 \\ -2 \\ -5 \end{pmatrix}$ and $\begin{pmatrix} 4 \\ -23 \\ 10 \end{pmatrix}$

(c) $\begin{pmatrix} 2.4 \\ 1.2 \\ 3.6 \\ 5.4 \end{pmatrix}$ and $\begin{pmatrix} 1 \\ -2 \\ -1 \\ 2 \end{pmatrix}$

(d) $\begin{pmatrix} 2\sqrt{3} \\ 1 \\ -5 \\ -4 \end{pmatrix}$ and $\begin{pmatrix} \sqrt{3} \\ 4 \\ \sqrt{0.36} \\ 2 \end{pmatrix}$

4.8. Use the geometrical definition of the inner product to calculate the angle between the following pairs of vectors

(a) $\begin{pmatrix} 1 \\ 0 \end{pmatrix}$ and $\begin{pmatrix} 1 \\ 4 \end{pmatrix}$

(b) $\begin{pmatrix} 1 \\ 1 \end{pmatrix}$ and $\begin{pmatrix} -1 \\ -1 \end{pmatrix}$

(c) $\begin{pmatrix} 1 \\ 2 \end{pmatrix}$ and $\begin{pmatrix} 3 \\ 6 \end{pmatrix}$

(d) $\begin{pmatrix} -1 \\ 3 \end{pmatrix}$ and $\begin{pmatrix} -4 \\ 1 \end{pmatrix}$

(e) $\begin{pmatrix} 2 \\ 1 \end{pmatrix}$ and $\begin{pmatrix} -1 \\ 3 \end{pmatrix}$

4.9. Determine the definition of the following planes

(a) with normal vector $\begin{pmatrix} 1 \\ -2 \\ 3 \end{pmatrix}$ through the origin

(b) with normal vector $\begin{pmatrix} -1.5 \\ 0.5 \\ 2 \end{pmatrix}$ through the point $\begin{pmatrix} 2 \\ 4 \\ -3 \end{pmatrix}$

(continued)

4.10. Determine, for each of the planes defined in Exercise 4.9. two other points that lie in the plane.

4.11. Calculate the work performed by an object moving in the presence of the forces in the figure:

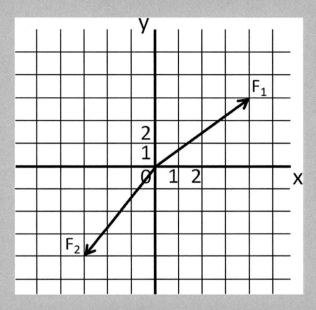

(a) for an object moving from the origin to the point (3,0) in the presence of force F_1 of 10N
(b) for an object moving from the origin to the point (0,−4) in the presence of force F_2 of 15N

4.2.2.2 Cross Product

Another way of multiplying two vectors is the cross product. There are some important differences between the inner product and the cross product. Whereas the inner product is defined in any space—two-, three- or higher dimensional—the cross product is only defined in 3D. And whereas the inner product results in a scalar (number), the cross product results in a vector that is perpendicular to the plane spanned by the two original vectors.

For the cross product, it is easiest to begin with its algebraic definition:

$$\vec{a} \times \vec{b} = \begin{pmatrix} a_1 \\ a_2 \\ a_3 \end{pmatrix} \times \begin{pmatrix} b_1 \\ b_2 \\ b_3 \end{pmatrix} = \begin{pmatrix} a_2 b_3 - a_3 b_2 \\ a_3 b_1 - a_1 b_3 \\ a_1 b_2 - a_2 b_1 \end{pmatrix}$$

This may look pretty weird, but will hopefully become clearer when we get along. First, let's focus on the calculation itself. To remember what you need to do when calculating a cross product it is important to realize that there is some logic in the calculation. To calculate the first element, you ignore the first elements of the original vectors and calculate the

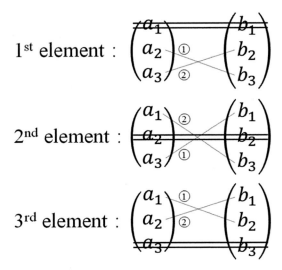

$$1^{st} \text{ element} : \begin{pmatrix} a_1 \\ a_2 \\ a_3 \end{pmatrix} \times \begin{pmatrix} b_1 \\ b_2 \\ b_3 \end{pmatrix}$$

$$2^{nd} \text{ element} : \begin{pmatrix} a_1 \\ a_2 \\ a_3 \end{pmatrix} \times \begin{pmatrix} b_1 \\ b_2 \\ b_3 \end{pmatrix}$$

$$3^{rd} \text{ element} : \begin{pmatrix} a_1 \\ a_2 \\ a_3 \end{pmatrix} \times \begin{pmatrix} b_1 \\ b_2 \\ b_3 \end{pmatrix}$$

Fig. 4.10 Illustration of the steps to take when calculating a cross product.

difference of the 'cross-wise' products of the remaining (second and third) elements of the vectors (Fig. 4.10). You do the same for the second and third elements of the cross product vector, but, and this is tricky, for the second element, you reverse the order. So instead of calculating $a_1 b_3 - a_3 b_1$, which would be the logical continuation of what I just explained you need to do to obtain the first element, you calculate the negative of that.

Example 4.2

Calculate the cross product of the vectors $\begin{pmatrix} 7 \\ 3 \\ -2 \end{pmatrix}$ and $\begin{pmatrix} 2 \\ -5 \\ 8 \end{pmatrix}$.

Following the algebraic definition of the cross product above, we find that the cross product of these two vectors is equal to:

$$\begin{pmatrix} 7 \\ 3 \\ -2 \end{pmatrix} \times \begin{pmatrix} 2 \\ -5 \\ 8 \end{pmatrix} = \begin{pmatrix} 3 \cdot 8 - (-2 \cdot -5) \\ -(7 \cdot 8 - (-2 \cdot 2)) \\ 7 \cdot -5 - 3 \cdot 2 \end{pmatrix} = \begin{pmatrix} 24 - 10 \\ -(56 + 4) \\ -35 - 6 \end{pmatrix} = \begin{pmatrix} 14 \\ -60 \\ -41 \end{pmatrix}$$

The steps taken in the calculation of the cross product of the vectors $\vec{a}$ and $\vec{b}$ are actually equivalent to calculating the determinant of a matrix which is explained in detail in Sect. 5.3.1. To know in which direction the cross product vector points with respect to the two original vectors, the right hand rule can be used. When you make a shape of your right hand as if you are pointing a gun at someone, while keeping your middle finger straight, your index finger oriented along the first vector and your middle finger oriented along the second vector, your thumb will point in the direction of the cross product (Fig. 4.11).

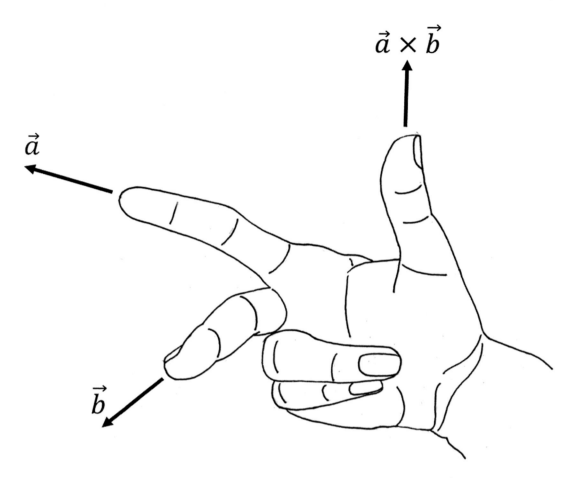

Fig. 4.11 Illustration of the right hand rule for calculating the cross product $\vec{a} \times \vec{b}$.

As can be understood from both the algebraic definition as well as from the right hand rule, the cross product is anti-commutative, i.e. $\vec{a} \times \vec{b} = -\vec{b} \times \vec{a}$.

There is also a geometric definition of the cross product:

$$\vec{a} \times \vec{b} = |\vec{a}| |\vec{b}| \sin \varphi \, \vec{n}$$

Similar to the inner product, $0 < \varphi < \pi$ is the angle between the two vectors when they originate at the same position (Fig. 4.7). The vector $\vec{n}$ is a unit vector; its orientation is determined by the right hand rule. Thus, the length of the cross product is equal to:

$$\left| \vec{a} \times \vec{b} \right| = |\vec{a}| |\vec{b}| \sin \varphi$$

This expression helps to gain an intuitive understanding of the cross product. Similar to what we did to obtain an intuitive understanding of the inner product in Sect. 4.2.2.1,

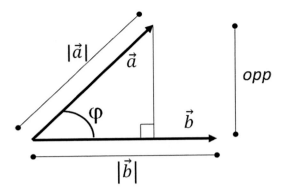

Fig. 4.12 Illustration supporting the intuitive understanding of the cross product of two vectors $\vec{a}$ and $\vec{b}$.

remember that the sine can be calculated as the ratio between the opposite edge and the hypothenuse in a right-angled triangle (Sect. 3.2). Hence, in the triangle drawn in Fig. 4.12, $\sin\varphi$ equals the part of $\vec{a}$ perpendicular to $\vec{b}$ (the opposite edge *opp*), divided by $|\vec{a}|$.

This makes the length of the cross product equal to:

$\left|\vec{a} \times \vec{b}\right| = |\vec{a}||\vec{b}| \sin\varphi = |\vec{a}||\vec{b}|\frac{opp}{|\vec{a}|} = |\vec{b}|opp$, hence to the part of $\vec{a}$ perpendicular to $\vec{b}$,

times the length of $\vec{b}$. Thus, the length of the cross product tells us how perpendicular two vectors are, with the maximum value of $|\vec{a}||\vec{b}|$ being attained when the two vectors are orthogonal. We can also use this exercise to understand the meaning of the length of the cross product in a more geometrical way. Reconsider Fig. 4.12 and the parallelogram spanned by the vectors $\vec{a}$ and $\vec{b}$. Its area is determined by $|\vec{b}|h$, where h is the height of the parallelogram.

But also $h = |\vec{a}| \sin\varphi = opp$. Hence, the area of the parallelogram is equal to $|\vec{b}|opp$. Therefore, the area of a parallelogram spanned by two vectors is equal to the length of the cross product of the two vectors.

There are many branches of physics in which the cross product is encountered. For example, torque, which expresses how much a force acting on an object makes that object rotate, is the cross product of the distance vector $\vec{r}$ originating at the axis of rotation and ending at the point where the force acts and the applied force vector $\vec{F}$:

$$T = \vec{r} \times \vec{F}$$

The length of $\vec{r}$ is known as the moment arm.

Another example from physics originates in the field of electromagnetism. Actually, probably the first time you encountered the right hand rule was in physics class, where you learned that it allows you to find the direction of the magnetic field due to a current in a coil or in a straight wire. In physics there are many applications of the right hand rule and several of them are related to cross products (see e.g. https://en.wikipedia.org/wiki/Right-hand_rule). One such example from electromagnetism is the force of a magnetic field acting on a moving charged particle. That force equals the product of the charge of the particle Q (a scalar) and the cross product of the velocity $\vec{v}$ and the magnetic field $\vec{B}$:

$$F = Q(\vec{v} \times \vec{B})$$

Exercises

4.12. Calculate the cross product of these pairs of vectors algebraically

(a) $\begin{pmatrix} 1 \\ 1 \\ 2 \end{pmatrix}$ and $\begin{pmatrix} 2 \\ 3 \\ 4 \end{pmatrix}$

(b) $\begin{pmatrix} 1 \\ 0 \\ -2 \end{pmatrix}$ and $\begin{pmatrix} -5 \\ 2 \\ 7 \end{pmatrix}$

(c) $\begin{pmatrix} -1 \\ -1 \\ -1 \end{pmatrix}$ and $\begin{pmatrix} 0.2 \\ 0.3 \\ 0.4 \end{pmatrix}$

4.13. Determine the plane

(a) spanned by the vectors $\begin{pmatrix} 1 \\ 1 \\ 1 \end{pmatrix}$ and $\begin{pmatrix} 2 \\ 3 \\ 4 \end{pmatrix}$ and going through the origin

(b) spanned by the vectors $\begin{pmatrix} 2 \\ 1 \\ -1 \end{pmatrix}$ and $\begin{pmatrix} 2 \\ 3 \\ 4 \end{pmatrix}$ and going through the point $\begin{pmatrix} 1 \\ 1 \\ 1 \end{pmatrix}$

4.3   Other Mathematical Concepts Related to Vectors

Now that the basics of vector addition, subtraction, scalar multiplication and multiplication hold no more secrets for you, it is time to get into more interesting mathematical concepts that rest on vector calculus and the relationships between pairs or sets of vectors.

4.3.1 Orthogonality, Linear Dependence and Correlation

In Sect. 4.2.2.1 we already stated that two vectors are orthogonal iff their inner product equals zero. In 2D or 3D this is equal to the two vectors being perpendicular to each other. In contrast, if two vectors have the same orientation and only differ in their length, they are *linearly dependent*. Formally, this can be defined as: two vectors $\vec{a}$ and $\vec{b}$ are linearly dependent if there is a scalar $s \neq 0$ such that $s\vec{a} = \vec{b}$. The definition of linear dependence can be extended to multiple vectors: n vectors $\vec{a}_1 \ldots \vec{a}_n$ are linearly dependent if there are scalars $s_1 \ldots s_n \neq 0$ such that $\sum_i s_i \vec{a}_i = 0$, or similarly, if one of the vectors can be expressed as a *linear combination* of the others. So far, so good. But what does correlation have to do with orthogonality and linear dependence? To understand this, we first consider correlation. Here, we define the correlation between two random variables X and Y by Pearson's correlation coefficient (or the Pearson product-moment correlation coefficient, or simply 'the correlation coefficient') as the *covariance* (cov) of the two variables divided by the product of their standard deviations σ:

$$\rho(X, Y) = \frac{\text{cov}(X, Y)}{\sigma_X \sigma_Y}$$

This is the correlation coefficient for a population. To estimate the population correlation coefficient ρ, you have to take a sample from the population and calculate the sample correlation coefficient r. Suppose that (x_1, y_1) ... (x_n, y_n) are n pairs of points sampled from X and Y of which the sample means of X and Y have been subtracted (X and Y have been *demeaned* or centered). Then the sample correlation coefficient r can be calculated as follows:

$$r(X, Y) = \frac{\frac{1}{n-1} \sum_{i=1}^{n} x_i y_i}{\sqrt{\frac{1}{n-1} \sum_{i=1}^{n} x_i^2} \sqrt{\frac{1}{n-1} \sum_{i=1}^{n} y_i^2}} = \frac{\sum_{i=1}^{n} x_i y_i}{\sqrt{\sum_{i=1}^{n} x_i^2} \sqrt{\sum_{i=1}^{n} y_i^2}} = \frac{\vec{x} \cdot \vec{y}}{|\vec{x}||\vec{y}|} = \cos \varphi$$

Hence, the sample correlation coefficient can be calculated by taking the inner product of two vectors representing pairs of demeaned variables and by dividing this product by the product of the vector norms. A graphical interpretation is the angle between them: the more the two (n-dimensional) vectors are oriented in the same direction, the more correlated the variables are. Thus, orthogonality denotes that the raw variables are perpendicular whereas uncorrelatedness denotes that the demeaned variables are perpendicular. To finalize this section we visualize the relations between orthogonality, linear independence

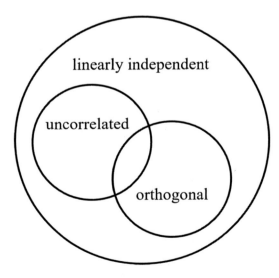

Fig. 4.13 Venn diagram illustrating the relations between orthogonality, linear independence and correlation.

and correlation in the Venn diagram in Fig. 4.13. This shows that all orthogonal vectors are independent, all uncorrelated vectors are independent and some uncorrelated vectors are orthogonal.

Note that when sampling in real life, zero correlation is very unlikely to occur: most data will show some correlation.

Exercises

4.14. Are the following pairs of vectors (1) independent?, (2) correlated?, (3) orthogonal?

(a) $\begin{pmatrix} 1 \\ 1 \\ 2 \\ 3 \end{pmatrix}$ and $\begin{pmatrix} 2 \\ 3 \\ 4 \\ 5 \end{pmatrix}$

(b) $\begin{pmatrix} 0 \\ 0 \\ 1 \\ 1 \end{pmatrix}$ and $\begin{pmatrix} 1 \\ 0 \\ 1 \\ 0 \end{pmatrix}$

(c) $\begin{pmatrix} 1 \\ -5 \\ 3 \\ -1 \end{pmatrix}$ and $\begin{pmatrix} 5 \\ 1 \\ 1 \\ 3 \end{pmatrix}$

(d) $\begin{pmatrix} -1 \\ -1 \\ 1 \\ 1 \end{pmatrix}$ and $\begin{pmatrix} 1 \\ -1 \\ 1 \\ -1 \end{pmatrix}$

(e) $\begin{pmatrix} 1 \\ 2 \\ 3 \\ 4 \end{pmatrix}$ and $\begin{pmatrix} 3 \\ 6 \\ 9 \\ 12 \end{pmatrix}$

4.3.2 Projection and Orthogonalization

Another interesting relation that rests on vector calculus is that between the inner product and the projection of one vector onto another. In Sect. 4.2.2.1, we derived that the inner product is equal to:

$$\vec{a} \cdot \vec{b} = |\vec{a}||\vec{b}| \cos \varphi = |\vec{a}||\vec{b}|\frac{adj}{|\vec{a}|} = |\vec{b}|adj, \text{ hence to the part of } \vec{a} \text{ in the direction of } \vec{b},$$

times the length of $\vec{b}$. The part of $\vec{a}$ in the direction of $\vec{b}$ (*adj*) is already very close to the projection of $\vec{a}$ on $\vec{b}$ as can be seen in Fig. 4.8, the only thing we need to do is to multiply it by a vector of unit length in the direction of $\vec{b}$, i.e. by $\vec{b}/|\vec{b}|$. Thus the projection of $\vec{a}$ on $\vec{b}$ is given by:

$$\frac{\vec{a} \cdot \vec{b}}{|\vec{b}|^2} \vec{b}.$$

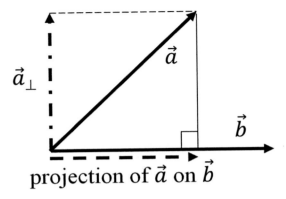

projection of $\vec{a}$ on $\vec{b}$

Fig. 4.14 Illustration of projection of vector $\vec{a}$ on vector $\vec{b}$ (*dashed vector*) and orthogonalized vector $\vec{a}_\perp$ (*dot-dashed vector*).

Once we know how to project two vectors onto each other, we can easily derive how to *orthogonalize* vector $\vec{a}$ with respect to vector $\vec{b}$; simply by subtracting the projection of vector $\vec{a}$ on vector $\vec{b}$ from vector $\vec{a}$ (as illustrated in Fig. 4.14):

$$\vec{a}_\perp = \vec{a} - \frac{\vec{a} \cdot \vec{b}}{\left|\vec{b}\right|^2} \vec{b}.$$

This procedure can be repeated to orthogonalize multiple vectors with respect to each other, e.g. three vectors $\vec{a}_1$, $\vec{a}_2$ and $\vec{a}_3$:

$$\vec{a}_{1,\perp} = \vec{a}_1$$
$$\vec{a}_{2,\perp} = \vec{a}_2 - \frac{\vec{a}_2 \cdot \vec{a}_{1,\perp}}{\left|\vec{a}_{1,\perp}\right|^2} \vec{a}_{1,\perp}$$
$$\vec{a}_{3,\perp} = \vec{a}_3 - \frac{\vec{a}_3 \cdot \vec{a}_{1,\perp}}{\left|\vec{a}_{1,\perp}\right|^2} \vec{a}_{1,\perp} - \frac{\vec{a}_3 \cdot \vec{a}_{2,\perp}}{\left|\vec{a}_{2,\perp}\right|^2} \vec{a}_{2,\perp}.$$

This process is known as Gram-Schmidt (or Schmidt-Gram) orthogonalization and can be extended to more vectors in a similar manner. Orthogonalization has some very useful applications, for example in *multiple linear regression*. Linear regression is a method to model the relationship between a scalar dependent variable (such as height) by an explanatory variable (such as age) by a linear model (i.e. a straight line in this simple example, see Fig. 4.15).

If there are more explanatory variables the method is referred to as multiple linear regression. Thus, describing this in a more mathematical way, in multiple linear regression one tries to predict a dependent random variable Y from m explanatory variables X_1 to X_m. To estimate the model you need n observations or measurements (Y_i, X_{1i}, ..., X_{mi}), $i = 1, \ldots, n$, for example from multiple people (if e.g. predicting some human property) or from multiple *voxels* (if e.g. predicting brain activation using functional magnetic

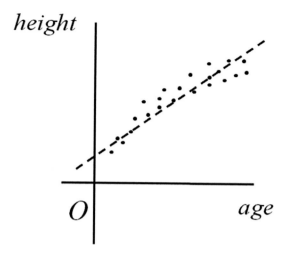

Fig. 4.15 Illustration of simple linear regression to predict height from age. *Dots* represent individual measurements, the *dashed line* is the linear model.

resonance imaging). Of course, the linear relationship will not be perfect, so there will be a deviation ε_i from the ith observation to the estimated model. Thus for the ith observation the relationship can be described as:

$$Y_i = \beta_1 X_{i1} + \beta_2 X_{i2} + \ldots + \beta_m X_{im} + \varepsilon_i$$

where $\beta_1 \ldots \beta_m$ are the model weights that need to be estimated. When putting these relationships together for all i the following matrix equation results:

$$\vec{Y} = X\vec{\beta} + \vec{\varepsilon}$$

where

$$X = \begin{pmatrix} X_{11} & \cdots & X_{1m} \\ X_{21} & \cdots & X_{2m} \\ \vdots & \ddots & \vdots \\ X_{n1} & \cdots & X_{nm} \end{pmatrix}$$

Thus, the columns of **X** contain the n observed values for each of the m explanatory variables. **X** is often referred to as the design matrix, its column vectors as *regressors*, $\vec{\beta}$ as the (vector of) regression coefficients and $\vec{\varepsilon}$ as the (vector of) error terms or noise. I here have already introduced the concept of a *matrix*, which will be explained in much more detail in the next Chapter. Let's consider an example of multiple linear regression to clarify its use.

Example 4.3

In science it is common practice to try to use one or more measured variables to predict another. One way to do this is with linear regression. For this example, let's suppose we want to know whether the average number of sugar-sweetened beverages taken per day and age can predict blood pressure (inspired by Nguyen et al. 2009). Suppose five participants are assessed; their data are displayed in the table below:

Participant number	# Beverages/day	Age (years)	Systolic blood pressure (mmHg)
1	5	20	140
2	3	25	130
3	7	50	150
4	1	40	120
5	0	39	110

To see whether the number of sweet beverages and age can predict systolic blood pressure, we can derive a linear equation per participant i of the form $BP_i = \beta_1 \cdot bev_i + \beta_2 \cdot age_i$. Here, BP is the systolic blood pressure and bev is the number of beverages per day. Since we have more equations (5) than unknowns (2), β_1 and β_2 can most likely not be determined such that the equations will be true for all participants. So instead of trying to solve a system of linear equations (see Sect. 2.4), we employ multiple linear regression and add an error term ε_i to each equation: $BP_i = \beta_1 \cdot bev_i + \beta_2 \cdot age_i + \varepsilon_i$. Thus:

$$140 = \beta_1 \cdot 5 + \beta_2 \cdot 20 + \varepsilon_1$$
$$130 = \beta_1 \cdot 3 + \beta_2 \cdot 25 + \varepsilon_2$$
$$150 = \beta_1 \cdot 7 + \beta_2 \cdot 50 + \varepsilon_3$$
$$120 = \beta_1 \cdot 1 + \beta_2 \cdot 40 + \varepsilon_4$$
$$110 = \beta_1 \cdot 0 + \beta_2 \cdot 39 + \varepsilon_5$$

We can use statistical programs like SPSS to apply multiple linear regression. In this case, after entering the participant details in SPSS and running multiple linear regression, we find that $\beta_1 = 5.498$ and $\beta_2 = 0.042$. In addition SPSS tells us that β_1 is significantly different from 0, whereas β_2 is not. This means that—in this example—the number of beverages does explain blood pressure for the participants, while age does not.

An important prerequisite for successful (multiple) linear regression is that the regressors are not *multicollinear*. Multicollinearity means that two or more of the regressors are highly correlated or that one of the regressors is (almost) a linear combination of the others (dependent). In such a case it is not possible to determine the regression coefficients reliably; a small change in the observations may result in a large change in the regression coefficients. A more phenomenological explanation is that it is not possible for the regression algorithm to decide where to 'put the weights'; at the regressor that is (almost) a linear combination of others or at those other regressors. Gram-Schmidt orthogonalization is one of the ways to resolve this problem, as orthogonal regressors are linearly independent (see Fig. 4.13). The next example will show this.

Example 4.4

A couple of years ago, we used Gram-Schmidt orthogonalization to allow the coupling of muscle activity (as measured by *electromyography* (EMG)) to brain activity (as measured by functional magnetic resonance imaging (fMRI); van Rootselaar et al. 2007, 2008) to gain further understanding of the brain areas that are involved in movement disorders such as tremor (Broersma et al. 2015). In the study on tremor we asked essential tremor patients (see also Sect. 3.4.2) to maintain a posture with extended arm and hand which evokes the tremor, and to alternate this with rest, while they were lying on their back in the magnetic resonance scanner. As one of the regressors for predicting brain activation over time in each voxel ($\vec{Y}$ in the multiple linear regression model above), we used the block design regressor of the task. This is a vector with ones for each scan during which patients maintained posture and zeros for each scan during which patients rested (giving one of the columns in the design matrix $\mathbf{X}$ in the multiple linear regression model above; see Fig. 4.16, top). During this experiment we also recorded EMG activity from the fore arm muscles as a measure of tremor intensity (see Fig. 4.16, middle). As tremor was evoked by posture, the EMG regressor—consisting of the average EMG intensity for each scan—was highly correlated with the block design regressor (compare Fig. 4.16 top and middle). Thus, if we would have simply put the block design regressor and the EMG regressor as two columns in the design matrix $\mathbf{X}$ in the linear regression model above to predict brain activation, the two regressors would have been dependent and we would not have been able to find reliable brain activations related to tremor. However, by first Gram-Schmidt orthogonalizing the EMG regressor with respect to the block design regressor and putting the result (see Fig. 4.16, bottom) in the design matrix instead of the EMG regressor, we were. In this orthogonalized EMG regressor we preserve only the tremor activity that is stronger or weaker than the average tremor activity, providing us with information about tremor variability. The brain areas for which activation correlates with this orthogonalized EMG regressor then presumably are involved in modulating tremor intensity. With this approach we found that the cerebellum is implicated in essential tremor (Broersma et al. 2015).

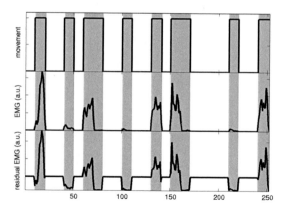

Fig. 4.16 Visualization of a block design regressor (movement; *top*), EMG regressor (*middle*) and result of Gram-Schmidt orthogonalization of the two (*bottom*). The final design matrix used for analyzing the fMRI data of the tremor experiment contained the block design regressor (*top*) as well as the orthogonalized EMG-regressor (*bottom*). Note that vector element values are plotted as a function of scan number here.

Glossary

Algebraically Using an expression in which only mathematical symbols and arithmetic operations are used.

Associativity Result is the same independent of how numbers are grouped (true for addition and multiplication of more than two numbers; see Box 4.1).

Commutativity Result is the same independent of the order of the numbers (true for addition and multiplication; see Box 4.1).

Coordinates Pairs (2D space) or triplets (2D space) of ordered numbers that determine a point's position in space.

Covariance A measure of joint variability of two random variables; if the two variables jointly increase and jointly decrease the covariance is positive, if one variable increases when the other decreases and vice versa, the covariance is negative.

Demeaned A set of variables from which the mean has been subtracted, i.e. with zero mean (also known as centered).

Determinant Property of a square matrix, which can be thought of as a scaling factor of the transformation the matrix represents (see also Chap. 5).

Distributivity Result is the same for multiplication of a number by a group of numbers added together, and each multiplication done separately after which all multiplications are added (in this case the algebraic expression is really much easier to understand than the words; see Box 4.1).

Electromyography Measurement of electrical muscle activity by (surface or needle) electrodes.

Element As in 'vector element': one of the entries in a vector.

Geometrically Using geometry and its methods and principles.

Geometry The branch of mathematics dealing with shapes, sizes and other spatial features.

iff 'If and only if': math lingo for a relation that holds in two directions, e.g. a iff b means 'if a then b *and* if b then a'.

Linear combination The result of multiplying a set of terms with constants and adding them.

Linear dependence Two vectors having the same orientation, differing only in their length.

Matrix A rectangular array of (usually) numbers.

Multicollinearity Two or more of a set of vectors or regressors are highly correlated or one of the vectors or regressors in this set is (almost) a linear combination of the others.

Multiple linear regression Predicting a dependent variable from a set of independent variables according to a linear model.

Norm Length or size of a vector.

Normal vector A vector that is perpendicular to a plane.

Null vector The result of subtracting a vector from itself.

Orthogonal Two vectors that are geometrically perpendicular or equally, have an inner product of zero.

Orthogonalize To make (two vectors) orthogonal.

Parallelepiped 3D figure formed by six parallelograms (it relates to a parallelogram like a cube relates to a square).

Parallelogram A quadrilateral with two pairs of parallel sides (note that a square is a parallelogram, but not every parallelogram is square).

Regressor An independent variable that can explain a dependent variable in a regression model, represented by a vector in the regression model

Scalar Here: a real number that can be used for scalar multiplication in which a vector multiplied by a scalar results in another vector.

Unit vector A vector divided by its own length, resulting in a vector with length one.

Vector Entity with a magnitude and direction, often (geometrically) indicated by an arrow.

Voxel Here: a sample of brain activation as measured on a 3D rectangular grid.

Work Product of the force exerted on an object and its displacement.

Symbols Used in This Chapter (in Order of Their Appearance)

$\vec{\cdot}$	Vector
$\vec{i}, \vec{j}, \vec{k}$	Vectors of length 1 along the principal x-, y-, and z-axes
$\lvert\vec{\cdot}\rvert$	Vector norm
ϕ	Here: angle between two vectors
$\cdot$	Inner product
$\cos$	Cosine
adj	Edge in a right-angled triangle adjacent to the acute angle
$\vec{n}$	Normal vector
$\times$	Here: cross product
$\sin$	Sine
opp	Edge in a right-angled triangle opposite to the acute angle cross product
cov	Covariance
σ	Standard deviation
$\vec{\cdot}_{\perp}$	Orthogonalized vector

Overview of Equations, Rules and Theorems for Easy Reference

Addition and subtraction of vectors

$$\vec{a} + \vec{b} + \vec{c} = \begin{pmatrix} a_1 \\ \vdots \\ a_n \end{pmatrix} + \begin{pmatrix} b_1 \\ \vdots \\ b_n \end{pmatrix} + \begin{pmatrix} c_1 \\ \vdots \\ c_n \end{pmatrix} = \begin{pmatrix} a_1 + b_1 + c_1 \\ \vdots \\ a_n + b_n + c_n \end{pmatrix}$$

Magnitude (length, module, absolute value, norm) of a vector

$$\lvert \vec{a} \rvert = \left\lvert \begin{pmatrix} a_1 \\ \vdots \\ a_n \end{pmatrix} \right\rvert = \sqrt{a_1^2 + \cdots + a_n^2}$$

Inner vector product

$$\vec{a} \cdot \vec{b} = \begin{pmatrix} a_1 \\ \vdots \\ a_n \end{pmatrix} \cdot \begin{pmatrix} b_1 \\ \vdots \\ b_n \end{pmatrix} = a_1 b_1 + \cdots a_n b_n \quad \text{or}$$

$$\vec{a} \cdot \vec{b} = |\vec{a}| |\vec{b}| \cos \varphi,$$

where $0 < \varphi < \pi$ is the angle between $\vec{a}$ and $\vec{b}$ when they originate at the same position.

Definition of a plane through the origin

$$\vec{n} \cdot \vec{x} = 0,$$

where $\vec{n}$ is a normal vector, perpendicular to the plane and $\vec{x}$ is any vector such that $\vec{n} \cdot \vec{x} = 0$

Cross vector product

$$\vec{a} \times \vec{b} = \begin{pmatrix} a_1 \\ a_2 \\ a_3 \end{pmatrix} \times \begin{pmatrix} b_1 \\ b_2 \\ b_3 \end{pmatrix} = \begin{pmatrix} a_2 b_3 - a_3 b_2 \\ a_3 b_1 - a_1 b_3 \\ a_1 b_2 - a_2 b_1 \end{pmatrix} \quad \text{or}$$

$$\vec{a} \times \vec{b} = |\vec{a}| |\vec{b}| \sin \varphi \, \vec{n}$$

where $0 < \varphi < \pi$ is the angle between $\vec{a}$ and $\vec{b}$ when they originate at the same position and $\vec{n}$ is a unit vector.

Correlation coefficient
The sample correlation between two random demeaned variables X and Y as defined by Pearson's correlation coefficient is given by:

$$r(X, Y) = \frac{\frac{1}{n-1} \sum_{i=1}^{n} x_i y_i}{\sqrt{\frac{1}{n-1} \sum_{i=1}^{n} x_i^2} \sqrt{\frac{1}{n-1} \sum_{i=1}^{n} y_i^2}} = \frac{\sum_{i=1}^{n} x_i y_i}{\sqrt{\sum_{i=1}^{n} x_i^2} \sqrt{\sum_{i=1}^{n} y_i^2}} = \frac{\vec{x} \cdot \vec{y}}{|\vec{x}| |\vec{y}|} = \cos \varphi$$

where $\vec{x}$ and $\vec{y}$ are two vectors representing the pairs of variables and $0 < \varphi < \pi$ is the angle between these vectors when they originate at the same position.

Projection
The projection of $\vec{a}$ on $\vec{b}$ is given by:

$$\frac{\vec{a} \cdot \vec{b}}{\left|\vec{b}\right|^2} \vec{b}$$

<u>(Gram-Schmidt) orthogonalization</u>
The orthogonalization of two vectors $\vec{a}_1$ and $\vec{a}_2$ or the Gram-Schmidt orthogonalization of three vectors $\vec{a}_1$, $\vec{a}_2$ and $\vec{a}_3$ is given by:

$$\vec{a}_{1,\perp} = \vec{a}_1$$

$$\vec{a}_{2,\perp} = \vec{a}_2 - \frac{\vec{a}_2 \cdot \vec{a}_{1,\perp}}{\left|\vec{a}_{1,\perp}\right|^2} \vec{a}_{1,\perp}$$

$$\vec{a}_{3,\perp} = \vec{a}_3 - \frac{\vec{a}_3 \cdot \vec{a}_{1,\perp}}{\left|\vec{a}_{1,\perp}\right|^2} \vec{a}_{1,\perp} - \frac{\vec{a}_3 \cdot \vec{a}_{2,\perp}}{\left|\vec{a}_{2,\perp}\right|^2} \vec{a}_{2,\perp}.$$

Answers to Exercises

4.1. The sum and difference of the two vectors are:

(a) $\begin{pmatrix} 9 \\ 7 \end{pmatrix}$ and $\begin{pmatrix} 5 \\ -1 \end{pmatrix}$

(b) $\begin{pmatrix} 1 \\ 3 \end{pmatrix}$ and $\begin{pmatrix} -5 \\ 31 \end{pmatrix}$

(c) $\begin{pmatrix} 5 \\ -21 \\ 10 \end{pmatrix}$ and $\begin{pmatrix} -3 \\ 25 \\ -4 \end{pmatrix}$

(d) $\begin{pmatrix} 3.6 \\ 4.8 \\ 9 \\ 7.8 \end{pmatrix}$ and $\begin{pmatrix} 1.2 \\ -2.4 \\ -1.8 \\ 3 \end{pmatrix}$

4.2. The vector $\begin{pmatrix} 7 \\ 3 \end{pmatrix}$ is indicated with a dashed arrow and the vector $\begin{pmatrix} 2 \\ 4 \end{pmatrix}$ is indicated with a dotted arrow, both starting at the origin. The geometrical sum of the two vectors is indicated in the left figure and the geometrical difference of the two vectors is indicated in the right figure, both by a drawn arrow. Some additional lines and copies of the (negative of the) arrows are indicated to visualize the supporting parallelogram.

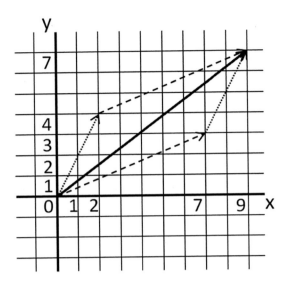
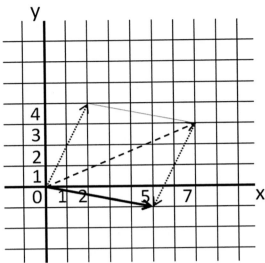

4.3. (a) $\begin{pmatrix} -2 \\ 3 \\ -21 \end{pmatrix}$

(b) $\begin{pmatrix} 3.5 \\ 0 \\ 2.9 \end{pmatrix}$

4.4. $P = (-2,-2)$, $Q = (-3,4)$ and $R = (1,2)$.

(a) Below, the points P, Q and R and their connecting vectors have been drawn.

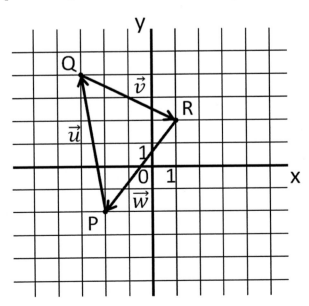

(b) $\vec{u} = \begin{pmatrix} -1 \\ 6 \end{pmatrix}$, $\vec{v} = \begin{pmatrix} 4 \\ -2 \end{pmatrix}$ and $\vec{w} = \begin{pmatrix} -3 \\ -4 \end{pmatrix}$

(c) The null vector (after connecting—i.e. adding—all vectors, we end up in the same location).

4.5. (a) One minute is 1/60th hour, thus the location of the airplane after 1 min is the velocity vector divided by 60: (3.33, 3, 1.66).

(b) 100 m is 0.1 km. The airplane climbs 100 km in 1 h, thus 0.1 km in 1/1000 h = 3.6 s.

(c) It takes 3.6 s to climb 100 m, thus $3.6 \times 10 \times 9 = 324$ s to climb to 9 km.

4.6. The inner product of each of these pairs of vectors is

(a) 0
(b) 0
(c) 15
(d) 4
(e) 1

4.7. The inner product of each of these pairs of vectors is

(a) 35
(b) 0
(c) 7.2
(d) −1

4.8. Use that $\vec{a} \cdot \vec{b} = |\vec{a}| |\vec{b}| \cos \varphi$.

(a) $\cos \varphi = 1/\sqrt{17}$, thus $\varphi \approx 76°$
(b) $\cos \varphi = -2/2 = -1$, thus $\varphi = 180°$
(c) $\cos \varphi = 15/\sqrt{5} \cdot \sqrt{45} = 15/\sqrt{225} = 15/15 = 1$, thus $\varphi = 0°$
(d) $\cos \varphi = 7/\sqrt{10} \cdot \sqrt{17}$, thus $\varphi \approx 58°$
(e) $\cos \varphi = 1/\sqrt{5} \cdot \sqrt{10}$, thus $\varphi \approx 82°$

4.9. (a) $x - 2y + 3z = 0$

(b) In general, a plane with normal vector $\begin{pmatrix} -1.5 \\ 0.5 \\ 2 \end{pmatrix}$ is given by $-1.5x + 0.5y + 2z = C$.

To find C, we substitute the point $\begin{pmatrix} 2 \\ 4 \\ -3 \end{pmatrix}$ that is in the plane: $-1.5 \cdot 2 + 0.5 \cdot 4 - 3 \cdot 2 = -3 + 2 - 6 = -7$ and thus this plane is given by $-1.5x + 0.5y + 2z = -7$.

4.10. (a) any combination of x, y and z that satisfies $x - 2y + 3z = 0$ is a point in the plane, e.g. $(-1, 1, 1)$ or $(2, 2, 2/3)$.

(b) similar to 4.10(a), $(1, 1, -3)$ is a point in this plane, but also $(0, 0, -3.5)$.

4.11. The work performed by the object is determined by the formula $work = F \cdot d \cdot \cos(\alpha)$, where F is the force (in N), d the displacement (in m) and α the angle between the displacement and the force.

(a) work $= 10 \cdot 3 \cdot 4/5 = 90/5 = 18$

(b) work $= 15 \cdot 4 \cdot 4/5 = 180/5 = 36$

4.12. (a) $\begin{pmatrix} 1 \\ 1 \\ 2 \end{pmatrix} \times \begin{pmatrix} 2 \\ 3 \\ 4 \end{pmatrix} = \begin{pmatrix} 4 - 6 \\ -(4 - 4) \\ 3 - 2 \end{pmatrix} = \begin{pmatrix} -2 \\ 0 \\ 1 \end{pmatrix}$

(b) $\begin{pmatrix} 1 \\ 0 \\ -2 \end{pmatrix} \times \begin{pmatrix} -5 \\ 2 \\ 7 \end{pmatrix} = \begin{pmatrix} 0 + 4 \\ -(7 - 10) \\ 2 - 0 \end{pmatrix} = \begin{pmatrix} 4 \\ 3 \\ 2 \end{pmatrix}$

(c) $\begin{pmatrix} -1 \\ -1 \\ -1 \end{pmatrix} \times \begin{pmatrix} 0.2 \\ 0.3 \\ 0.4 \end{pmatrix} = \begin{pmatrix} -0.4 + 0.3 \\ -(-0.4 + 0.2) \\ -0.3 + 0.2 \end{pmatrix} = \begin{pmatrix} -0.1 \\ 0.2 \\ -0.1 \end{pmatrix}$

4.13. Remember that a plane is determined by its normal vector and that a normal vector can be determined by the cross product of two points (vectors) in the plane.

(a) the normal vector for this plane is given by

$$\begin{pmatrix} 1 \\ 1 \\ 1 \end{pmatrix} \times \begin{pmatrix} 2 \\ 3 \\ 4 \end{pmatrix} = \begin{pmatrix} 4 - 3 \\ -(4 - 2) \\ 3 - 2 \end{pmatrix} = \begin{pmatrix} 1 \\ -2 \\ 1 \end{pmatrix},$$

thus the plane (through the origin) is $x - 2y + z = 0$.

(b) the normal vector for this plane is given by

$$\begin{pmatrix} 2 \\ 1 \\ -1 \end{pmatrix} \times \begin{pmatrix} 2 \\ 3 \\ 4 \end{pmatrix} = \begin{pmatrix} 4 + 3 \\ -(8 + 2) \\ 6 - 2 \end{pmatrix} = \begin{pmatrix} 7 \\ -10 \\ 4 \end{pmatrix},$$

Thus the plane (through the point $(1, 1, 1)$) is $7x - 10y + 4z = 1$.

4.14. Tip: remember that two vectors $\vec{a}$ and $\vec{b}$ are dependent if there is a scalar s such that $s\vec{a} = \vec{b}$, correlated if $(\vec{a} - \bar{a}) \cdot (\vec{b} - \bar{b}) = 0$ (where $\bar{}$ denotes the mean over all vector elements) and orthogonal if $\vec{a} \cdot \vec{b} = 0$

(a) independent, correlated and not orthogonal

(b) independent, uncorrelated and not orthogonal

(c) independent, correlated and orthogonal
(d) independent, uncorrelated and orthogonal
(e) dependent, correlated and not orthogonal

References

Online Sources of Information: History

http://www.math.mcgill.ca/labute/courses/133f03/VectorHistory.html

Online Sources of Information: Methods

http://www.britannica.com/topic/vector-mathematics
http://mathinsight.org/vector_introduction
https://www.khanacademy.org/math/linear-algebra/vectors_and_spaces/
https://en.wikipedia.org/wiki/Linear_independence
https://en.wikipedia.org/wiki/Multicollinearity
https://en.wikipedia.org/wiki/Linear_regression

Papers

M. Broersma, A.M. van der Stouwe, A.W. Buijink, B.M. de Jong, P.F. Groot, J.D. Speelman, M.A. Tijssen, A.F. van Rootselaar, N.M. Maurits, Bilateral cerebellar activation in unilaterally challenged essential tremor. Neuroimage Clin. **11**, 1–9 (2015)

A.F. van Rootselaar, R. Renken, B.M. de Jong, J.M. Hoogduin, M.A. Tijssen, N.M. Maurits, fMRI analysis for motor paradigms using EMG-based designs: a validation study. Hum. Brain Mapp. **28**(11), 1117–1127 (2007)

A.F. van Rootselaar, N.M. Maurits, R. Renken, J.H. Koelman, J.M. Hoogduin, K.L. Leenders, M.A. Tijssen, Simultaneous EMG-functional MRI recordings can directly relate hyperkinetic movements to brain activity. Hum. Brain Mapp. **29**(12), 1430–1441 (2008)

S. Nguyen, H.K. Choi, R.H. Lustig, C. Hsu, Sugar Sweetened Beverages, Serum Uric Acid, and Blood Pressure in Adolescents. The Journal of Pediatrics, **154**(6), 807–813 (2009) http://doi.org/10.1016/j.jpeds.2009.01.015

5

Matrices

Natasha Maurits

After reading this chapter you know:

- what matrices are and how they can be used,
- how to perform addition, subtraction and multiplication of matrices,
- that matrices represent linear transformations,
- the most common special matrices,
- how to calculate the determinant, inverse and eigendecomposition of a matrix, and
- what the decomposition methods SVD, PCA and ICA are, how they are related and how they can be applied.

5.1 What Are Matrices and How Are They Used?

Matrices, in mathematics, are rectangular arrays of (usually) numbers. Their entries are called *elements* and can also be symbols or even expressions. Here, we discuss matrices of numbers. Of course, these numbers can be of any type, such as integer, real or complex (see Sect. 1.2). For most practical applications, the matrix elements have specific meanings, such as the distance between cities, demographic measures such as survival probabilities (represented in a *Leslie matrix*) or the absence or presence of a path between nodes in a *graph* (represented in an *adjacency matrix*), which is applied in *network theory*. Network theory has seen a surge of interest in recent years because of its wide applicability in the study of e.g. social communities, the world wide web, the brain, regulatory relationships between genes, metabolic pathways and logistics. We here first consider the simple example of roads between the four cities in Fig. 5.1.

N. Maurits (✉)
Department of Neurology, University Medical Center Groningen, Groningen, The Netherlands
e-mail: n.m.maurits@umcg.nl

N. Maurits, B. Ćurčić-Blake, *Math for Scientists*, DOI 10.1007/978-3-319-57354-0_5

A matrix **M** describing the distances between the cities is given by:

$$M = \begin{bmatrix} 0 & 22 & 14 & 10 \\ 22 & 0 & 8 & 12 \\ 14 & 8 & 0 & 13 \\ 10 & 12 & 13 & 0 \end{bmatrix}$$

Here, each row corresponds to a 'departure' city A–D and each column to an 'arrival' city A–D. For example, the distance from city B to city C (second row, third column) is 8, as is the distance from city C to city B (third row, second column). Cities A and B are not connected directly, but can be reached through cities C or D. In both cases, the distance between cities A and B amounts to 22 (14 + 8 or 12 + 10; first row, second column and second row, first column).

One of the advantages of using a matrix instead of a table is that they can be much more easily manipulated by computers in large-scale calculations. For example, matrices are used to store all relevant data for weather predictions and for predictions of air flow patterns around newly designed *airfoils*.

Historically, matrices, which were first known as 'arrays', have been used to solve systems of linear equations (see Chap. 2 and Sect. 5.3.1) for centuries, even dating back to 1000 years BC in China. The name 'matrix', however, wasn't introduced until the nineteenth century, when James Joseph Sylvester thought of a matrix as an entity that can give rise to smaller matrices by removing columns and/or rows, as if they are 'born' from the womb of a common parent. Note that the word 'matrix' is related to the Latin word for mother: 'mater'.

By removing columns or rows and considering just one row or one column of a matrix, we obtain so-called row- or column-matrices, which are simply row- or column-vectors, of

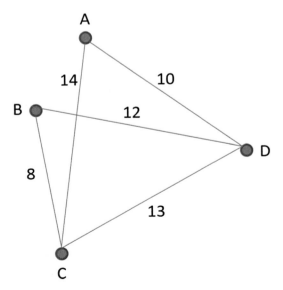

Fig. 5.1 Four cities A, B, C and D and the distances between them. In this example city A cannot be reached from city B directly, but can be reached via city C or city D.

course. Hence, vectors (see Chap. 4) are just special forms of matrices. Typically, matrices are indicated by capital letters, often in bold (as in this book) and sometimes, when writing manually, by underlined or overlined capital letters. The *order* or size of a matrix is always indicated by first mentioning the number of rows and then the number of columns. Thus, an $m \times n$ matrix has m rows and n columns. For example, $A = \begin{pmatrix} 1 & 6 & -3 \\ 0.5 & 7 & 4 \end{pmatrix}$ is a 2×3 matrix. When the number of rows equals the number of columns, as in the distance example above, the matrix is called *square*. An element of a matrix $\mathbf{A}$ in its ith row and j^{th} column can be indicated by a_{ij}, $a_{i,j}$ or $a(i,j)$. Thus, in the example matrix $\mathbf{A}$ above $a_{1,2} = 6$ and $a_{2,1} = 0.5$. This is very similar to how vector elements are indicated, except that vector elements have one index instead of two.

5.2 Matrix Operations

All common mathematical operations, such as addition, subtraction and multiplication, can be executed on matrices, very similar to how they are executed on vectors (see Sect. 4.2.1).

5.2.1 Matrix Addition, Subtraction and Scalar Multiplication

Since the concepts of vector addition, subtraction and scalar multiplication should by now be familiar to you, explaining the same operations for matrices becomes quite easy. The geometrical definition that could be used for vectors is not available for matrices, so we here limit the definition of these operations to the algebraic one. The notation for matrix elements that was just introduced in the previous section helps to define these basic operations on matrices. For example, addition of matrices $\mathbf{A}$ and $\mathbf{B}$ (of the same size) is defined as:

$$(A + B)_{ij} = a_{ij} + b_{ij}$$

Thus, the element at position (i,j) in the sum matrix is obtained by adding the elements at the same position in the matrices $\mathbf{A}$ and $\mathbf{B}$. Subtraction of matrices $\mathbf{A}$ and $\mathbf{B}$ is defined similarly as:

$$(A - B)_{ij} = a_{ij} - b_{ij},$$

and multiplication of a matrix $\mathbf{A}$ by a scalar s is defined as:

$$(sA)_{ij} = sa_{ij}$$

Exercises

5.1. In the figure below three cities A, B and C are indicated with two roads between them.

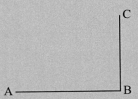

There is no direct connection between cities A and C. The distance matrix between the three cities is:

$$\begin{pmatrix} 0 & 12 & 21 \\ 12 & 0 & 9 \\ 21 & 9 & 0 \end{pmatrix}$$

(a) Copy the figure and add the distances to the roads.
(b) A direct (straight) road between cities A and C is built: how does the distance matrix change?
(c) There are also three public parks D, E and F in the area. The (shortest) distance matrix between the cities (rows) and the parks (columns) is given by:

$$\begin{pmatrix} 15 & 24 & 0 \\ 9 & 12 & 12 \\ 0 & 15 & 15 \end{pmatrix}$$

Indicate the location of the parks in the figure, assuming that the new road between A and C has been built.

5.2. Add and subtract (i.e. first-second matrix) the following pairs of matrices:

(a) $\begin{pmatrix} 3 & 4 \\ -1 & 8 \end{pmatrix}$ and $\begin{pmatrix} 2 & -2 \\ 3 & 7 \end{pmatrix}$

(b) $\begin{pmatrix} 3 & -7 & 4 \\ -2 & 6 & 5 \\ 1 & -2 & -9 \end{pmatrix}$ and $\begin{pmatrix} 4 & 3 & 2 \\ 1 & -2 & -4 \\ -5 & 8 & 11 \end{pmatrix}$

(c) $\begin{pmatrix} 1.2 & 3.2 & -1.5 \\ 3.4 & 2.3 & -3.2 \end{pmatrix}$ and $\begin{pmatrix} 0.8 & -1.6 & 0.5 \\ 1.7 & -1.3 & 1.2 \end{pmatrix}$

5.3. Calculate

(a) $\begin{pmatrix} 1 & 2 \\ 7 & 1 \\ 1 & 4 \end{pmatrix} + 3 \begin{pmatrix} 1 & 0 \\ 0 & 1 \\ 1 & 0 \end{pmatrix} - 2 \begin{pmatrix} -1 & 1 \\ 0 & 4 \\ -2 & -3 \end{pmatrix}$

(b) $-\begin{pmatrix} 0.5 & 3.1 \\ 6.7 & 2.4 \end{pmatrix} - 2 \begin{pmatrix} 1 & 0.7 \\ 1.2 & 0.7 \end{pmatrix} + 1.5 \begin{pmatrix} 4 & 3 \\ 8 & 3 \end{pmatrix}$

5.2.2 Matrix Multiplication and Matrices as Transformations

Matrix addition, subtraction and scalar multiplication are quite straightforward generalizations of the same operations on vectors. For multiplication, the story is a little different, although there is a close relation between matrix multiplication and the vector inner product. A matrix product is only well defined if the number of columns in the first matrix equals the number of rows in the second matrix. Suppose $\mathbf{A}$ is an $m \times n$ matrix and $\mathbf{B}$ is an $n \times p$ matrix, then the product $\mathbf{AB}$ is defined by:

$$(AB)_{ij} = \sum_{k=1}^{n} a_{ik}b_{kj}$$

If you look closer at this definition, multiplying two matrices turns out to entail repeated vector inner product calculations, e.g. the element in the third row and second column of the product is the inner product of the third row of the first matrix with the second column of the second matrix. Let's work this out for one abstract and for one concrete example:

$$\begin{pmatrix} a_{11} & a_{12} & a_{13} \\ a_{21} & a_{22} & a_{23} \end{pmatrix} \begin{pmatrix} b_{11} & b_{12} \\ b_{21} & b_{22} \\ b_{31} & b_{32} \end{pmatrix} = \begin{pmatrix} a_{11}b_{11}+a_{12}b_{21}+a_{13}b_{31} & a_{11}b_{12}+a_{12}b_{22}+a_{13}b_{32} \\ a_{21}b_{11}+a_{22}b_{21}+a_{23}b_{31} & a_{21}b_{12}+a_{22}b_{22}+a_{23}b_{32} \end{pmatrix}$$

$$\begin{pmatrix} 3 & 4 & 2 \\ -2 & -1 & 0 \\ 2 & -3 & 7 \end{pmatrix} \begin{pmatrix} 1 & 2 \\ 3 & 4 \\ 5 & 6 \end{pmatrix} = \begin{pmatrix} 3\cdot1+4\cdot3+2\cdot5 & 3\cdot2+4\cdot4+2\cdot6 \\ -2\cdot1-1\cdot3+0\cdot5 & -2\cdot2-1\cdot4+0\cdot6 \\ 2\cdot1-3\cdot3+7\cdot5 & 2\cdot2-3\cdot4+7\cdot6 \end{pmatrix} = \begin{pmatrix} 25 & 34 \\ -5 & -8 \\ 28 & 34 \end{pmatrix}$$

Exercises

5.4. Which pairs of the following matrices could theoretically be multiplied with each other and in which order? Write down all possible products.
   A with order 2×3
   B with order 3×4
   C with order 3×3
   D with order 4×2

5.5. What is the order (size) of the resulting matrix when multiplying two matrices of orders
   (a) 2×3 and 3×7
   (b) 2×2 and 2×1
   (c) 1×9 and 9×1

5.6. Calculate whichever product is possible: **AB**, **BA**, none or both.

   (a) $\mathbf{A} = \begin{pmatrix} 3 & 4 \\ -1 & 8 \end{pmatrix}$, $\mathbf{B} = \begin{pmatrix} 2 & -2 \\ 3 & 7 \end{pmatrix}$

   (b) $\mathbf{A} = \begin{pmatrix} 3 & -7 \\ -2 & 6 \\ 1 & -2 \end{pmatrix}$, $\mathbf{B} = \begin{pmatrix} 4 & 3 & 2 \\ 1 & -2 & -4 \\ -5 & 8 & 11 \end{pmatrix}$

   (c) $\mathbf{A} = \begin{pmatrix} 1.2 & 3.2 & -1.5 \\ 3.4 & 2.3 & -3.2 \end{pmatrix}$, $\mathbf{B} = \begin{pmatrix} 0.8 & -1.6 & 0.5 \\ 1.7 & -1.3 & 1.2 \end{pmatrix}$

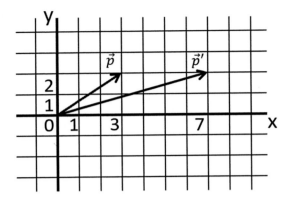

Fig. 5.2 Illustration of how vector $\vec{p}$ is transformed into vector $\vec{p}'$ by matrix M (see main text).

So far, we discussed matrix operations as operations on abstract entities. However, matrices are actually commonly used to represent (linear) *transformations*. What this entails can be easily demonstrated by showing the effect of the application of matrices on vectors. Application here means that the matrix and vector are multiplied. Let's first illustrate this by an example. Consider a vector $\vec{p} = \begin{pmatrix} 3 \\ 2 \end{pmatrix}$, a matrix $M = \begin{pmatrix} 1 & 2 \\ 2 & -2 \end{pmatrix}$ and its product

$$\vec{p}' = M\vec{p} = \begin{pmatrix} 1 & 2 \\ 2 & -2 \end{pmatrix} \begin{pmatrix} 3 \\ 2 \end{pmatrix} = \begin{pmatrix} 7 \\ 2 \end{pmatrix} \text{ (see Fig. 5.2).}$$

First, observe that when we calculate $M\vec{p}$ another 2×1 vector results, since we multiply a 2×2 matrix and a 2×1 vector. Apparently, judging from Fig. 5.2, applying the matrix **M** to the vector $\vec{p}$ transforms the vector; it rotates it and changes its length. To understand this transformation, let's see how this matrix transforms the two basis vectors of 2D space; $\begin{pmatrix} 1 \\ 0 \end{pmatrix}$ and $\begin{pmatrix} 0 \\ 1 \end{pmatrix}$. These two vectors are called *basis vectors*, because any vector $\begin{pmatrix} a \\ b \end{pmatrix}$ in 2D space can be built from them by a linear combination as follows: $\begin{pmatrix} a \\ b \end{pmatrix} = a \begin{pmatrix} 1 \\ 0 \end{pmatrix} + b \begin{pmatrix} 0 \\ 1 \end{pmatrix}$. The matrix **M** transforms the first basis vector $\begin{pmatrix} 1 \\ 0 \end{pmatrix}$ to $\begin{pmatrix} 1 \\ 2 \end{pmatrix}$ and the second basis vector $\begin{pmatrix} 0 \\ 1 \end{pmatrix}$ to $\begin{pmatrix} 2 \\ -2 \end{pmatrix}$, i.e. the two columns of the matrix **M** show how the transform affects the basis vectors. The effect of **M** on any vector $\begin{pmatrix} a \\ b \end{pmatrix}$ is thus equal to $M \begin{pmatrix} a \\ b \end{pmatrix} = a \begin{pmatrix} 1 \\ 2 \end{pmatrix} + b \begin{pmatrix} 2 \\ -2 \end{pmatrix}$. For the example in Fig. 5.2 this results indeed in $M \begin{pmatrix} 3 \\ 2 \end{pmatrix} = 3 \begin{pmatrix} 1 \\ 2 \end{pmatrix} + 2 \begin{pmatrix} 2 \\ -2 \end{pmatrix} = \begin{pmatrix} 7 \\ 2 \end{pmatrix}$.

There are some special geometric transformation matrices, both in 2D, as well as 3D. One that is often encountered is the rotation matrix, that leaves the length of a vector unaltered, but rotates it around the origin in 2D or 3D. The rotation matrix has applications in e.g., the study of rigid body movements and in the manipulation of images, as in the preprocessing of functional magnetic resonance imaging (fMRI) scans. The transformation matrix that rotates a vector around the origin (in 2D) over an angle θ (counter clockwise) is given by $\begin{pmatrix} \cos\theta & -\sin\theta \\ \sin\theta & \cos\theta \end{pmatrix}$, as illustrated in Fig. 5.3 (cf. Chap. 3 on the geometrical definition of sine and cosine).

Another common geometric transformation matrix is the *shearing* matrix, that transforms a square in 2D into a parallelogram (see Fig. 5.4). Applying the matrix transformation $\begin{pmatrix} 1 & k \\ 0 & 1 \end{pmatrix}$ results in shearing along the x-axis (y-coordinate remains unchanged), whereas applying the matrix transformation $\begin{pmatrix} 1 & 0 \\ k & 1 \end{pmatrix}$ results in shearing along the y-axis (x-coordinate remains unchanged). This transformation matrix is also applied in e.g., the preprocessing of fMRI scans.

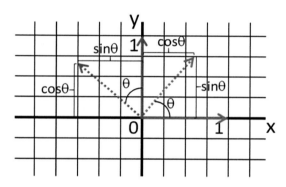

Fig. 5.3 Illustration of how the 2D basis vectors, $\begin{pmatrix} 1 \\ 0 \end{pmatrix}$ in *red* and $\begin{pmatrix} 0 \\ 1 \end{pmatrix}$ in *blue*, are affected by a counter-clockwise rotation over an angle θ. The vectors before rotation are indicated by *drawn arrows*, the vectors after rotation are indicated by *dotted arrows*.

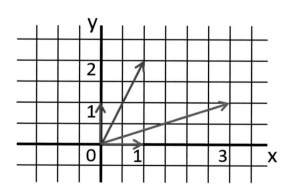

Fig. 5.4 Illustration of how the two 2D basis vectors are affected by shearing along the x- and y-axes. The applied transformation matrix for shearing along the x-axis is $\begin{pmatrix} 1 & 3 \\ 0 & 1 \end{pmatrix}$ (results in *red*) and the applied transformation matrix for shearing along the y-axis is $\begin{pmatrix} 1 & 0 \\ 2 & 1 \end{pmatrix}$ (results in *blue*).

5.2.3 Alternative Matrix Multiplication

Just as there are different ways of multiplying vectors, there are also different ways of multiplying matrices, that are less common, however. For example, matrices can be multiplied element-wise; this product is also referred to as the *Hadamard product*, Schur product or pointwise product:

$$(A \circ B)_{ij} = a_{ij} b_{ij}$$

This only works when the two matrices have the same size. Another matrix product is the *Kronecker product*, which is a generalization of the tensor product (or dyadic product) for vectors. For vectors this product is equal to an $n \times 1$ column vector being multiplied by a $1 \times n$ row vector, which results in an $n \times n$ matrix when following the standard multiplication rules for matrices:

$$\vec{a} \otimes \vec{b} = \vec{a}\vec{b}^T = \begin{pmatrix} a_1 \\ a_2 \\ \vdots \\ a_n \end{pmatrix} (b_1 \quad b_2 \quad \cdots \quad b_n) = \begin{pmatrix} a_1 b_1 & a_1 b_2 & \cdots & a_1 b_n \\ a_2 b_1 & a_2 b_2 & \cdots & a_2 b_n \\ \vdots & \vdots & \ddots & \vdots \\ a_n b_1 & a_n b_2 & \cdots & a_n b_n \end{pmatrix}$$

For two matrices $\mathbf{A}$ $(m \times n)$ and $\mathbf{B}$ $(p \times q)$, the Kronecker product $\mathbf{A} \otimes \mathbf{B}$ is defined by:

$$A \otimes B = \begin{pmatrix} a_{11}B & a_{12}B & \cdots & a_{1n}B \\ a_{21}B & a_{22}B & \cdots & a_{2n}B \\ \vdots & \vdots & \ddots & \vdots \\ a_{m1}B & a_{m2}B & \cdots & a_{mn}B \end{pmatrix}$$

Note that the size of the matrices $\mathbf{A}$ and $\mathbf{B}$ does not need to match for the Kronecker product. The Kronecker product has proven its use in the study of matrix theory (linear algebra), e.g. in solving matrix equations such as the Sylvester equation $\mathbf{AX} + \mathbf{XB} = \mathbf{C}$ for general $\mathbf{A}$, $\mathbf{B}$ and $\mathbf{C}$ (see e.g. Horn and Johnson 1994).

Exercises

5.7. Calculate the Hadamard or pointwise product of

(a) $\begin{pmatrix} 1 & 2 \\ -1 & 1 \end{pmatrix}$ and $\begin{pmatrix} 2 & 3 \\ 4 & 1 \end{pmatrix}$

(b) $\begin{pmatrix} 2 & -0.3 & 1 \\ 1.5 & 7 & -0.4 \end{pmatrix}$ and $\begin{pmatrix} 1.4 & 9 & 0.5 \\ 8 & -0.1 & 10 \end{pmatrix}$

(continued)

5.8. Calculate the Kronecker product of

(a) $\begin{pmatrix} 1 & 2 \\ -1 & 1 \end{pmatrix}$ and $(3 \quad -1 \quad 4)$

(b) $(-2 \quad -3)$ and $\begin{pmatrix} 0 & 1 & 2 \\ 3 & 4 & 5 \\ 6 & 7 & 8 \end{pmatrix}$

5.2.4 Special Matrices and Other Basic Matrix Operations

There are some special forms of matrices and other basic operations on matrices that should be known before we can explain more interesting examples and applications of matrices.

The *identity* or *unit matrix* of size n is a square matrix of size $n \times n$ with ones on the diagonal and zeroes elsewhere. It is often referred to as $\mathbf{I}$ or, if necessary to explicitly mention its size, as $\mathbf{I_n}$. The identity matrix is a special *diagonal matrix*. More generally, a diagonal matrix only has non-zero elements on the diagonal and zeroes elsewhere. If a diagonal matrix is extended with non-zero elements only above or only below the diagonal, we speak about an upper or lower *triangular matrix*, respectively. A matrix that is symmetric around the diagonal, i.e. for which $a_{ij}=a_{ji}$, is called *symmetric*. An example of a symmetric matrix is a distance matrix, such as encountered in the very first example in this chapter. A matrix is *skew-symmetric* if $a_{ij}=-a_{ji}$. A matrix is called *sparse* if most of its elements are zero. In contrast, a matrix whose elements are almost all non-zero is called *dense*. A *logical* matrix (or binary or Boolean) matrix only contains zeroes and ones.

One operation that was already encountered in the previous chapter on vectors is the *transpose*. The transpose of an n × m matrix $\mathbf{A}$ is an m × n matrix of which the elements are defined as:

$$\left(A^T\right)_{ij} = a_{ji}$$

A generalization of the transpose to complex-valued elements is the *conjugate transpose*, for which the elements are defined by:

$$\left(A^*\right)_{ij} = \bar{a}_{ji},$$

where the overbar indicates the complex conjugate (see Sect. 1.2.4.1). For real matrices, the transpose is equal to the conjugate transpose. In quantum mechanics the conjugate transpose is indicated by † (dagger) instead of *. Notice that $(\mathbf{AB})^T = \mathbf{B}^T\mathbf{A}^T$ and $(\mathbf{AB})^* = \mathbf{B}^*\mathbf{A}^*$ for matrices $\mathbf{A}$ and $\mathbf{B}$ for which the matrix product $\mathbf{AB}$ is possible.

Now that we have introduced the conjugate transpose, we can also introduce a few more special (complex) matrices. For a *Hermitian* matrix: $\mathbf{A} = \mathbf{A}^*$, for a *normal* matrix: $\mathbf{A}^*\mathbf{A} = \mathbf{A}\mathbf{A}^*$ and for a *unitary* matrix: $\mathbf{A}\mathbf{A}^* = \mathbf{I}$. Note that all unitary and Hermitian matrices are normal, but the reverse is not true. The unitary matrices will return when we explain singular value decomposition in Sect. 5.3.3.

All special matrices that were introduced in this section are summarized in Table 5.1 with their abstract definition and a concrete example.

Table 5.1 Special matrices with their abstract definition and an example

	Definition	Example
Identity/unit	$a_{ij} = 0$ if $i \neq j$ $a_{ij} = 1$ if $i = j$	$\begin{pmatrix} 1 & 0 & 0 & 0 & 0 \\ 0 & 1 & 0 & 0 & 0 \\ 0 & 0 & 1 & 0 & 0 \\ 0 & 0 & 0 & 1 & 0 \\ 0 & 0 & 0 & 0 & 1 \end{pmatrix}$
Diagonal	$a_{ij} = 0$ if $i \neq j$ $a_{ij} \neq 0$ if $i = j$	$\begin{pmatrix} 2 & 0 & 0 \\ 0 & -3 & 0 \\ 0 & 0 & 7 \end{pmatrix}$
Lower-triangular	$a_{ij} = 0$ if $i < j$ $a_{ij} \neq 0$ if $i \geq j$	$\begin{pmatrix} 1 & 0 & 0 \\ 3 & -2 & 0 \\ -5 & 6 & -4 \end{pmatrix}$
Upper-triangular	$a_{ij} = 0$ if $i > j$ $a_{ij} \neq 0$ if $i \leq j$	$\begin{pmatrix} 1 & 3 & 19 \\ 0 & -2 & -8 \\ 0 & 0 & -4 \end{pmatrix}$
Symmetric	$a_{ij} = a_{ji}$	$\begin{pmatrix} 0 & 12 & 21 \\ 12 & 0 & -9 \\ 21 & -9 & 0 \end{pmatrix}$
Skew-symmetric	$a_{ij} = -a_{ji}$	$\begin{pmatrix} 0 & -12 & -21 \\ 12 & 0 & -9 \\ 21 & 9 & 0 \end{pmatrix}$
Sparse	Most elements are zero	$\begin{pmatrix} 1 & 0 & 0 & 0 & 0 \\ 0 & 2 & 0 & -3 & 0 \\ 0 & 0 & 0.5 & 0 & 0 \\ 0 & 0 & 0 & -1 & 0 \\ -1 & 0 & 0 & 0 & 7 \end{pmatrix}$
Dense	Most elements are non-zero	$\begin{pmatrix} 3 & 7 & -5 & 8 & -1 \\ 0 & 2 & 4 & -3 & 0 \\ 6 & 4 & -1 & 34 & 7 \\ 2 & 0 & 6 & -11 & 3 \\ -1 & -3 & 5 & 6 & 7 \end{pmatrix}$
Logical	$a_{ij} \in \{0,1\}$	$\begin{pmatrix} 0 & 0 & 1 & 0 & 1 \\ 0 & 1 & 0 & 1 & 0 \\ 1 & 1 & 1 & 0 & 0 \\ 0 & 1 & 0 & 1 & 0 \\ 1 & 1 & 0 & 0 & 1 \end{pmatrix}$
Hermitian	$\mathbf{A} = \mathbf{A}^*$	$\begin{pmatrix} 3 & 1+i & 8 \\ 1-i & 7 & -i \\ 8 & i & -2 \end{pmatrix}$
Normal	$\mathbf{A}^*\mathbf{A} = \mathbf{A}\mathbf{A}^*$	$\begin{pmatrix} 3 & 1+i & 8 \\ 1-i & 7 & -i \\ 8 & i & -2 \end{pmatrix}$
Unitary	$\mathbf{A}\mathbf{A}^* = \mathbf{I}$	$\frac{1}{2}\begin{pmatrix} 1+i & 1-i \\ 1-i & 1+i \end{pmatrix}$

Exercises

5.9. In what sense (according to the definitions in Table 5.1) are the following matrices special? Mention at least one property.

(a) $\begin{pmatrix} 1 & 1 & 0 \\ 1 & 1 & 0 \\ 0 & 0 & 1 \end{pmatrix}$

(b) $\begin{pmatrix} 1 & 0 & 3 \\ 0 & 0 & 0 \\ 0 & 0 & 1 \end{pmatrix}$

(c) $\begin{pmatrix} 0 & 4 & 3 \\ -4 & 0 & -7 \\ -3 & 7 & 0 \end{pmatrix}$

(d) $\begin{pmatrix} 1 & 4 & 3 \\ 0 & 1 & -7 \\ 0 & 0 & 1 \end{pmatrix}$

(e) $\begin{pmatrix} 1 & 0 & 0 \\ 0 & 23 & 0 \\ 0 & 0 & -7 \end{pmatrix}$

(f) $\begin{pmatrix} 1 & 0 & 0 \\ 0 & 1 & 0 \\ 0 & 0 & 1 \end{pmatrix}$

5.10. Determine the conjugate transpose of

(a) $\begin{pmatrix} 1 & 2 & 3 \\ -i & 1 & -3 - 2i \\ 5 & 4 + 5i & 3 \end{pmatrix}$

(b) $\begin{pmatrix} 1 & 2 & 3 \\ -1 & 1 & -3 \\ 5 & 4 & 0 \end{pmatrix}$

(c) $\begin{pmatrix} 4 & 0 & 3 - 2i \\ 19 + i & -3 & -3 \\ -8i & -11 - i & 17 \end{pmatrix}$

5.3 More Advanced Matrix Operations and Their Applications

Now that the basics of matrix addition, subtraction, scalar multiplication and multiplication hold no more secrets for you, it is time to get into more interesting mathematical concepts that rest on matrix calculus and to illustrate their applications.

5.3.1 Inverse and Determinant

One of the oldest applications of matrices is to solve systems of linear equations that were introduced to you in Chap. 2. Consider the following system of linear equations with three unknowns x, y and z:

$$3x + 4y - 2z = 5$$
$$-2x - 2y + z = -3$$
$$x + y - 7z = -18$$

This system can also be written as a matrix equation:

$$\begin{pmatrix} 3 & 4 & -2 \\ -2 & -2 & 1 \\ 1 & 1 & -7 \end{pmatrix} \begin{pmatrix} x \\ y \\ z \end{pmatrix} = \begin{pmatrix} 5 \\ -3 \\ -18 \end{pmatrix} \text{ or as}$$

$$M \begin{pmatrix} x \\ y \\ z \end{pmatrix} = \begin{pmatrix} 5 \\ -3 \\ -18 \end{pmatrix} \text{ where } M = \begin{pmatrix} 3 & 4 & -2 \\ -2 & -2 & 1 \\ 1 & 1 & -7 \end{pmatrix}$$

Such a system and more generally, systems of n equations in n unknowns, can be solved by using *determinants*, which is actually similar to using the *inverse* of a matrix to calculate the solution to a system of linear equations. The inverse of a square matrix $\mathbf{A}$ is the matrix $\mathbf{A}^{-1}$ such that $\mathbf{AA}^{-1} = \mathbf{A}^{-1}\mathbf{A} = \mathbf{I}$. The inverse of a matrix $\mathbf{A}$ does not always exist; if it does $\mathbf{A}$ is called invertible. Notice that $(\mathbf{AB})^{-1} = \mathbf{B}^{-1}\mathbf{A}^{-1}$ for square, invertible matrices $\mathbf{A}$ and $\mathbf{B}$. Now suppose that the matrix $\mathbf{M}$ in the matrix equation above has an inverse $\mathbf{M}^{-1}$. In that case, the solution to the equation would be given by:

$$\begin{pmatrix} x \\ y \\ z \end{pmatrix} = I \begin{pmatrix} x \\ y \\ z \end{pmatrix} = M^{-1}M \begin{pmatrix} x \\ y \\ z \end{pmatrix} = M^{-1} \begin{pmatrix} 5 \\ -3 \\ -18 \end{pmatrix}$$

Mathematically, the inverse of 2×2 matrix $\begin{pmatrix} a & b \\ c & d \end{pmatrix}$ is given by $\frac{1}{ad-bc} \begin{pmatrix} d & -b \\ -c & a \end{pmatrix}$ where $ad - bc$ is the determinant $\begin{vmatrix} a & b \\ c & d \end{vmatrix}$ of the matrix. This illustrates why a square matrix has an inverse if and only if (iff) its determinant is not zero, as division by zero would result in infinity. For a 3×3 matrix or higher-dimensional matrices the inverse can still be calculated by hand, but it quickly becomes cumbersome. In general, for a matrix $\mathbf{A}$:

$$A^{-1} = \frac{1}{\det(A)} adj(A),$$

where $adj(\mathbf{A})$ is the *adjoint* of $\mathbf{A}$. The adjoint of a matrix $\mathbf{A}$ is the transpose of the *cofactor matrix*. Each (i,j)-element of a cofactor matrix is given by the determinant of the matrix that remains when the i-th row and j-th column are removed, multiplied by -1 if $i + j$ is odd. In Box 5.1 the inverse is calculated of the matrix $\mathbf{M}$ given in the example that we started this section with.

Box 5.1 Example of calculating the inverse of a matrix

To be able to calculate $\mathbf{M}^{-1}$ where $M = \begin{pmatrix} 3 & 4 & -2 \\ -2 & -2 & 1 \\ 1 & 1 & -7 \end{pmatrix}$ we first need to know how to calculate the determinant of a 3×3 matrix. This is done by first choosing a reference row or column and calculating the cofactors for that row or column. Then the determinant is equal to the sum of the products of the elements of that row or column with its cofactors. This sounds rather abstract so let's calculate det($\mathbf{M}$) by taking its first row as a reference.

$$\det(M) = \det \begin{pmatrix} 3 & 4 & -2 \\ -2 & -2 & 1 \\ 1 & 1 & -7 \end{pmatrix}$$

$$= 3 \begin{vmatrix} -2 & 1 \\ 1 & -7 \end{vmatrix} - 4 \begin{vmatrix} -2 & 1 \\ 1 & -7 \end{vmatrix} - 2 \begin{vmatrix} -2 & -2 \\ 1 & 1 \end{vmatrix}$$

$$= 3 \cdot 13 - 4 \cdot 13 - 2 \cdot 0 = -13$$

For the matrix $\mathbf{M}$ its cofactor matrix is given by

$$\begin{pmatrix} \begin{vmatrix} -2 & 1 \\ 1 & -7 \end{vmatrix} & -\begin{vmatrix} -2 & 1 \\ 1 & -7 \end{vmatrix} & \begin{vmatrix} -2 & -2 \\ 1 & 1 \end{vmatrix} \\ -\begin{vmatrix} 4 & -2 \\ 1 & -7 \end{vmatrix} & \begin{vmatrix} 3 & -2 \\ 1 & -7 \end{vmatrix} & -\begin{vmatrix} 3 & 4 \\ 1 & 1 \end{vmatrix} \\ \begin{vmatrix} 4 & -2 \\ -2 & 1 \end{vmatrix} & -\begin{vmatrix} 3 & -2 \\ -2 & 1 \end{vmatrix} & \begin{vmatrix} 3 & 4 \\ -2 & -2 \end{vmatrix} \end{pmatrix} = \begin{pmatrix} 13 & -13 & 0 \\ 26 & -19 & 1 \\ 0 & 1 & 2 \end{pmatrix}.$$

Hence, the adjoint matrix of $\mathbf{M}$ is its transpose $\begin{pmatrix} 13 & 26 & 0 \\ -13 & -19 & 1 \\ 0 & 1 & 2 \end{pmatrix}$. Thus, the inverse of $\mathbf{M}$ is given by:

$$M^{-1} = \frac{1}{-13} \begin{pmatrix} 13 & 26 & 0 \\ -13 & -19 & 1 \\ 0 & 1 & 2 \end{pmatrix}$$

To verify that the inverse that we calculated in Box 5.1 is correct, it suffices to verify that the matrix multiplied by its inverse equals the identity matrix. Thus, for matrix $\mathbf{M}$ we verify that $\mathbf{M}^{-1}\mathbf{M} = \mathbf{I}$:

$$M^{-1}M = \frac{1}{-13} \begin{pmatrix} 13 & 26 & 0 \\ -13 & -19 & 1 \\ 0 & 1 & 2 \end{pmatrix} \begin{pmatrix} 3 & 4 & -2 \\ -2 & -2 & 1 \\ 1 & 1 & -7 \end{pmatrix}$$

$$= -\frac{1}{13} \begin{pmatrix} -13 & 0 & 0 \\ 0 & -13 & 0 \\ 0 & 0 & -13 \end{pmatrix} = \begin{pmatrix} 1 & 0 & 0 \\ 0 & 1 & 0 \\ 0 & 0 & 1 \end{pmatrix} = I$$

Finally, now that we have found $\mathbf{M}^{-1}$, we can find the solution to the system of equations that we started with:

$$\begin{pmatrix} x \\ y \\ z \end{pmatrix} = M^{-1} \begin{pmatrix} 5 \\ -3 \\ -18 \end{pmatrix} = -\frac{1}{13} \begin{pmatrix} 13 & 26 & 0 \\ -13 & -19 & 1 \\ 0 & 1 & 2 \end{pmatrix} \begin{pmatrix} 5 \\ -3 \\ -18 \end{pmatrix} = -\frac{1}{13} \begin{pmatrix} -13 \\ -26 \\ -39 \end{pmatrix} = \begin{pmatrix} 1 \\ 2 \\ 3 \end{pmatrix}$$

Finally, this solution can again be verified by inserting the solution in the system of equations that was given at the beginning of this section.

Cramer's rule is an explicit formulation of using determinants to solve systems of linear equations. We first formulate it for a system of three linear equations in three unknowns

$$a_1 x + b_1 y + c_1 z = d_1$$
$$a_2 x + b_2 y + c_2 z = d_2$$
$$a_3 x + b_3 y + c_3 z = d_3$$

Its associated determinant is:

$$D = \begin{vmatrix} a_1 & b_1 & c_1 \\ a_2 & b_2 & c_2 \\ a_3 & b_3 & c_3 \end{vmatrix}.$$

We can also define the determinant

$$D_x = \begin{vmatrix} d_1 & b_1 & c_1 \\ d_2 & b_2 & c_2 \\ d_3 & b_3 & c_3 \end{vmatrix},$$

which is the determinant of the system's associated matrix with its first column replaced by the vector of constants and similarly

$$D_y = \begin{vmatrix} a_1 & d_1 & c_1 \\ a_2 & d_2 & c_2 \\ a_3 & d_3 & c_3 \end{vmatrix} \quad \text{and} \quad D_z = \begin{vmatrix} a_1 & b_1 & d_1 \\ a_2 & b_2 & d_2 \\ a_3 & b_3 & d_3 \end{vmatrix}.$$

Then, x, y and z can be calculated as:

$$x = \frac{D_x}{D}, y = \frac{D_y}{D}, z = \frac{D_z}{D}.$$

Similarly, the solution of a system of n linear equations in n unknowns:

$$\begin{pmatrix} a_{11} & a_{12} & \cdots & a_{1n} \\ a_{21} & a_{22} & \cdots & a_{2n} \\ \vdots & \vdots & \ddots & \vdots \\ a_{n1} & a_{n2} & \cdots & a_{nn} \end{pmatrix} \begin{pmatrix} x_1 \\ x_2 \\ \vdots \\ x_n \end{pmatrix} = \begin{pmatrix} b_1 \\ b_2 \\ \vdots \\ b_n \end{pmatrix} \quad \text{or} \quad Ax = b,$$

according to Cramer's rule is given by:

$$x_1 = \frac{D_{x_1}}{D}, x_2 = \frac{D_{x_2}}{D}, \quad \ldots \quad , x_n = \frac{D_{x_n}}{D},$$ where D_{x_i} is the determinant of the matrix formed

by replacing the i-th column of $\mathbf{A}$ by the column vector $\vec{b}$. Note that Cramer's rule only applies when $D \neq 0$. Unfortunately, Cramer's rule is computationally very inefficient for larger systems and thus not often used in practice.

Exercises

5.11. Show that using Cramer's rule to find the solution to a system of two linear equations in two unknowns $ax + by = c$ and $dx + ey = f$ is the same as applying the inverse of the matrix $\begin{pmatrix} a & b \\ d & e \end{pmatrix}$ to the vector $\begin{pmatrix} c \\ f \end{pmatrix}$.

5.12. Find the solution to the following systems of linear equations using Cramer's rule:
   (a) (Example 2.6) $3x+5=5y \wedge 2x-5y=6$
   (b) $4x-2y-2z=10 \wedge 2x+8y+4z=32 \wedge 30x+12y-4z=24$

5.13. Find the solution to the following systems of linear equations using the matrix inverse:
   (a) (Exercise 2.4a) $x - 2y = 4 \wedge \frac{x}{3} - y = \frac{4}{3}$
   (b) (Example 2.7) $2x+y+z=4 \wedge x-7-2y=-3z \wedge 2y+10-2z=3x$

Now that you know how to calculate the determinant of a matrix, it is easy to recognize that the algebraic definition of the cross-product introduced in Sect. 4.2.2.2 is similar to calculating the determinant of a very special matrix:

$$\vec{a} \times \vec{b} = \begin{pmatrix} a_1 \\ a_2 \\ a_3 \end{pmatrix} \times \begin{pmatrix} b_1 \\ b_2 \\ b_3 \end{pmatrix} = \begin{pmatrix} a_2 b_3 - a_3 b_2 \\ a_3 b_1 - a_1 b_3 \\ a_1 b_2 - a_2 b_1 \end{pmatrix} = \begin{vmatrix} \vec{i} & \vec{j} & \vec{k} \\ a_1 & a_2 & a_3 \\ b_1 & b_2 & b_3 \end{vmatrix}$$

$$= \begin{vmatrix} a_2 & a_3 \\ b_2 & b_3 \end{vmatrix} \vec{i} - \begin{vmatrix} a_1 & a_3 \\ b_1 & b_3 \end{vmatrix} \vec{j} + \begin{vmatrix} a_1 & a_2 \\ b_1 & b_2 \end{vmatrix} \vec{k}$$

Finding the inverse of larger square matrices and thus finding the solution to larger systems of linear equations may also be accomplished by calculating the inverse matrix by hand. However, as you will surely have noticed when doing the exercises, finding the solution of a system of three linear equations with three unknowns calculating the inverse matrix by hand is already rather cumbersome, tedious and error-prone. Computers do a much better job than we at this sort of task which is why (numerical) mathematicians have developed clever, fast computer algorithms to determine matrix inverses. There is even a whole branch of numerical mathematics that focuses on solving systems of equations that can be represented by sparse matrices, as fast as possible (Saad 2003). The relevance of sparse matrices is that they naturally appear in many scientific or engineering applications whenever *partial differential equations* (see Box 5.2 and Chap. 6) are numerically solved on a grid. Typically, only local physical interactions play a role in such models of reality and thus only neighboring grid cells interact, resulting in sparse matrices. Examples are when heat dissipation around an underground gas tube or turbulence in the wake of an airfoil needs to be simulated. In Box 5.2 a simple example is worked out to explain how a *discretized* partial differential equation can result in a sparse matrix.

Box 5.2 How discretizing a partial differential equation can yield a sparse matrix

We here consider the *Laplace equation* for a *scalar function u* in two dimensions (indicated by x and y), which is e.g. encountered in the fields of electromagnetism and fluid dynamics:

$$\nabla^2 u = \frac{\partial^2 u}{\partial x^2} + \frac{\partial^2 u}{\partial y^2} = 0$$

It can describe airflow around airfoils or water waves, under specific conditions. In Sect. 6.11 partial derivatives are explained in more detail. To solve this equation on a rectangle, a grid or lattice of evenly spaced points (with distance h) can be overlaid as in Fig. 5.5.

One method to discretize the Laplace equation on this grid (see for details e.g. Press et al. n.d.) is:

$$\frac{u_{i-1,j} - 2u_{i,j} + u_{i+1,j}}{h^2} + \frac{u_{i,j-1} - 2u_{i,j} + u_{i,j+1}}{h^2} = 0$$

or

$$u_{ij} = \frac{1}{4}\left(u_{i-1,j} + u_{i+1,j} + u_{i,j-1} + u_{i,j+1}\right) \tag{5.1}$$

Here, i runs over the grid cells in x-direction and j runs over the grid cells in y-direction. This discretized equation already shows that the solution in a point (i,j) (in red in the figure) is only determined by the values in its local neighbours, i.e. the four points (i + 1,j), (i − 1,j), (i, j + 1) and (i, j − 1) (in blue in the figure) directly surrounding it. There are alternative discretizations that use variable spacings in the two grid directions and/or fewer or more points; the choice is determined by several problem parameters such as the given *boundary conditions*.

Here, I only want to illustrate how solving this problem involves a sparse matrix. Typically, iterative solution methods are used, meaning that the solution is approximated step-by-step starting from an initial solution and adapting the solution at every iteration until the solution hardly changes anymore. The solution at iteration step n + 1 (indicated by a superscript) can then be derived from the solution at time step n e.g. as follows (cf. Eq. 5.1):

$$u_{i,j}^{n+1} = \frac{1}{4}\left(u_{i-1,j}^n + u_{i+1,j}^n + u_{i,j-1}^n + u_{i,j+1}^n\right)$$

Here, we can recognize a (sparse) matrix transformation as I will explain now. Suppose we are dealing with the 11 × 7 grid in the figure. We can then arrange the values of u at iteration step n in a vector e.g. by row: $\vec{u}^n = \left(u_{1,1}^n \quad \cdots \quad u_{1,11}^n \quad u_{2,1}^n \quad \cdots \quad u_{2,11}^n \quad u_{3,1}^n \quad \cdots \quad u_{7,11}^n \right)^T$ and similarly for the values of u at iteration step n + 1. The 77 × 77 matrix transforming $\vec{u}^n$ into $\vec{u}^{n+1}$ then has only four non-zero entries and 73 zeroes in every row (with some exceptions for the treatment of the boundary grid points) and is thus sparse. It should be noted that this particular iterative solution method is slow and that much faster methods are available.

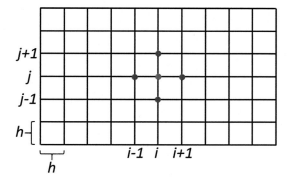

Fig. 5.5 Grid with equal spacing *h* for numerical solution of the Laplace equation in two dimensions. To solve for the point (*i,j*) (in *red*), values in the surrounding points (in *blue*) can be used.

5.3.2 Eigenvectors and Eigenvalues

A first mathematical concept that rests on matrix calculus and that is part of other interesting matrix decomposition methods (see Sect. 5.3.3) is that of *eigenvectors* and *eigenvalues*. An eigenvector $\vec{v}$ of a square matrix **M** is a non-zero column vector that does not change its orientation (although it may change its length by a factor λ) as a result of the transformation represented by the matrix. In mathematical terms:

$$M\vec{v} = \lambda\vec{v},$$

where λ is a scalar known as the eigenvalue. Eigenvectors and eigenvalues have many applications, of which we will encounter a few in the next section. For example, they allow to determine the principal axes of rotational movements of rigid bodies (dating back to eighteenth century Euler), to find common features in images as well as statistical data reduction.

To find the eigenvectors of a matrix **M**, the following system of linear equations has to be solved:

$$(M - \lambda I)\vec{v} = 0.$$

It is known that this equation has a solution iff the determinant of $M - \lambda I$ is zero. Thus, the eigenvalues of **M** are those values of λ that satisfy $|M - \lambda I| = 0$. This determinant is a polynomial in λ of which the highest power is the order of the matrix **M**. The eigenvalues can be found by finding the *roots* of this polynomial, known as the characteristic polynomial of **M**. There are as many roots as the order (highest power) of the polynomial. Once the eigenvalues are known, the related eigenvectors can be found by solving for $\vec{v}$ in $M\vec{v} = \lambda\vec{v}$. This probably sounds quite abstract, so a concrete example is given in Box 5.3.

Box 5.3 Example of calculating the eigenvalues and eigenvectors of a matrix

To find the eigenvalues of the matrix $\begin{pmatrix} 3 & 1 \\ 2 & 4 \end{pmatrix}$, we first determine its characteristic equation as the determinant

$$\left| \begin{pmatrix} 3 & 1 \\ 2 & 4 \end{pmatrix} - \lambda \begin{pmatrix} 1 & 0 \\ 0 & 1 \end{pmatrix} \right| = \left| \begin{pmatrix} 3-\lambda & 1 \\ 2 & 4-\lambda \end{pmatrix} \right| = (3-\lambda)(4-\lambda) - 2 = \lambda^2 - 7\lambda + 10$$

Since $\lambda^2 - 7\lambda + 10 = (\lambda - 2)(\lambda - 5)$, the roots of this polynomial are given by $\lambda = 2$ and $\lambda = 5$. The eigenvector for $\lambda = 2$ follows from:

$$\begin{pmatrix} 3 & 1 \\ 2 & 4 \end{pmatrix} \begin{pmatrix} x \\ y \end{pmatrix} = 2 \begin{pmatrix} x \\ y \end{pmatrix} \Rightarrow \begin{pmatrix} 3x+y \\ 2x+4y \end{pmatrix} = \begin{pmatrix} 2x \\ 2y \end{pmatrix}$$
$$\Rightarrow 3x + y = 2x \wedge 2x + 4y = 2y$$
$$\Rightarrow x + y = 0 \wedge 2x + 2y = 0$$
$$\Rightarrow x = -y$$
$$\Rightarrow \begin{pmatrix} x \\ y \end{pmatrix} = \begin{pmatrix} -1 \\ 1 \end{pmatrix}$$

Notice that $\begin{pmatrix} -1 \\ 1 \end{pmatrix}$ is the eigenvector in this case, because for any multiple of this vector it is true that $x = -y$.

Similarly, the eigenvector for $\lambda = 5$ follows from:

$$\begin{pmatrix} 3 & 1 \\ 2 & 4 \end{pmatrix} \begin{pmatrix} x \\ y \end{pmatrix} = 5 \begin{pmatrix} x \\ y \end{pmatrix} \Rightarrow \begin{pmatrix} 3x+y \\ 2x+4y \end{pmatrix} = \begin{pmatrix} 5x \\ 5y \end{pmatrix}$$
$$\Rightarrow 3x + y = 5x \wedge 2x + 4y = 5y$$
$$\Rightarrow -2x + y = 0 \wedge 2x - y = 0$$
$$\Rightarrow x = \frac{1}{2}y$$
$$\Rightarrow \begin{pmatrix} x \\ y \end{pmatrix} = \begin{pmatrix} 1 \\ 2 \end{pmatrix}$$

Notice that $\begin{pmatrix} 1 \\ 2 \end{pmatrix}$ is the eigenvector in this case, because for any multiple of this vector it is true that $2x = y$.

Hence, the matrix $\begin{pmatrix} 3 & 1 \\ 2 & 4 \end{pmatrix}$ has eigenvectors $\begin{pmatrix} -1 \\ 1 \end{pmatrix}$ and $\begin{pmatrix} 1 \\ 2 \end{pmatrix}$ with eigenvalues 2 and 5, respectively.

Exercises

5.14. Consider the shearing matrix $\begin{pmatrix} 1 & 2 \\ 0 & 1 \end{pmatrix}$. Without calculation, which are the eigenvectors of this matrix and why?

5.15. Calculate the eigenvectors and eigenvalues of

(continued)

(a) $\begin{pmatrix} 7 & 0 & 0 \\ 0 & -19 & 0 \\ 0 & 0 & 2 \end{pmatrix}$

(b) $\begin{pmatrix} 2 & 1 \\ -1 & 4 \end{pmatrix}$

(c) $\begin{pmatrix} 2 & 2 & -1 \\ 1 & 3 & -1 \\ 1 & 4 & -2 \end{pmatrix}$

(d) $\begin{pmatrix} 5 & 1 \\ 4 & 5 \end{pmatrix}$

5.3.3 Diagonalization, Singular Value Decomposition, Principal Component Analysis and Independent Component Analysis

In the previous section I already alluded to several applications using eigenvalues and eigenvectors. In this section I will discuss some of the most well-known and most often used methods that employ *eigendecomposition* and that underlie these applications.

The simplest is *diagonalization*. A matrix **M** can be diagonalized if it can be written as

$$M = VDV^{-1} \tag{5.2}$$

where **V** is an invertible matrix and **D** is a diagonal matrix. First note that **V** has to be square to be invertible, so that **M** also has to be square to be diagonalizable. Now let's find out how eigenvalues and eigenvectors play a role in Eq. 5.2. We can rewrite Eq. 5.2 by right multiplication with **V** to:

$$MV = VD$$

or similarly, when we indicate the columns of **V** by $\vec{v}_i$ and the elements of D by d_i, $i = 1, \ldots, n$, then:

$$M\begin{pmatrix} \vec{v}_1 & \vec{v}_2 & \cdots & \vec{v}_n \end{pmatrix} = \begin{pmatrix} \vec{v}_1 & \vec{v}_2 & \cdots & \vec{v}_n \end{pmatrix}D = \begin{pmatrix} d_1\vec{v}_1 & d_2\vec{v}_2 & \cdots & d_n\vec{v}_n \end{pmatrix}$$

This shows that in Eq. 5.2 the columns of **V** must be the eigenvectors of **M** and the diagonal elements of **D** must be the eigenvalues of **M**. We just noted that only square matrices can potentially be diagonalized. What *singular value decomposition* does is generalize the concept of diagonalization to any matrix, making it a very powerful method.

For an intuitive understanding of SVD you can think of a matrix as a large table of data, e.g. describing which books from a collection of 500 a 1000 people have read. In that case the matrix could contains ones for books (in 500 columns) that a particular reader (in 1000 rows) read and zeroes for the books that weren't read. SVD can help summarize the data in this large matrix. When SVD is applied to this particular matrix it will help identify specific sets of books that are often read by the same readers. For example, SVD may find that all thrillers in the collection are often read together, or all science fiction novels, or all romantic books; the composition of these subcollections will be expressed in the *singular vectors*, and their importance in the *singular values*. Now each reader's reading behavior can be expressed in a much more compact way than by just tabulating all books that he or she read. Instead, SVD allows to say e.g., that a reader is mostly interested in science fiction novels, or maybe that in addition, there is some interest in popular science books. Let's see how this compression of data can be achieved by SVD by first looking at its mathematics.

In *singular value decomposition* (SVD) an $m \times n$ rectangular matrix **M** is decomposed into a product of three matrices as follows:

$$M = U\Sigma V^*$$

where **U** is a unitary (see Table 5.1) $m \times m$ matrix, Σ an $m \times n$ diagonal matrix with non-negative real entries and **V** another unitary $n \times n$ matrix. To determine these matrices you have to calculate the sets of *orthonormal* eigenvectors of $\mathbf{MM}^*$ and $\mathbf{M}^*\mathbf{M}$. This can be done, as $\mathbf{MM}^*$ and $\mathbf{M}^*\mathbf{M}$ are square. The orthonormal eigenvectors of the former are the columns of **U** and the orthonormal eigenvectors of the latter are the columns of **V**. The square roots of the non-zero eigenvalues of $\mathbf{MM}^*$ or $\mathbf{M}^*\mathbf{M}$ (which do not differ) are the so-called *singular values* of M and form the diagonal elements of Σ in decreasing order, completed by zeroes if necessary. The columns of **U** are referred to as the *left-singular vectors* of **M** and the columns of **V** are referred to as the *right-singular vectors* of **M**. For an example which can also be intuitively understood, see Box 5.4.

Box 5.4 Example of SVD of a real square matrix and its intuitive understanding

To determine the SVD of the matrix $M = \begin{pmatrix} 1 & -2 \\ 2 & -1 \end{pmatrix}$, we first determine the eigenvalues and eigenvectors of the matrix $\mathbf{M}^*\mathbf{M}$ (which is equal to $\mathbf{M}^T\mathbf{M}$ in this real case) to get the singular values and right-singular vectors of **M**. Thus, we determine its characteristic equation as the determinant

$$\left| \begin{pmatrix} 1 & 2 \\ -2 & -1 \end{pmatrix} \begin{pmatrix} 1 & -2 \\ 2 & -1 \end{pmatrix} - \lambda \begin{pmatrix} 1 & 0 \\ 0 & 1 \end{pmatrix} \right| = \left| \begin{pmatrix} 5 & -4 \\ -4 & 5 \end{pmatrix} - \lambda \begin{pmatrix} 1 & 0 \\ 0 & 1 \end{pmatrix} \right| =$$

$$\left| \begin{pmatrix} 5-\lambda & -4 \\ -4 & 5-\lambda \end{pmatrix} \right| = (5-\lambda)^2 - 16 = \lambda^2 - 10\lambda + 9 = (\lambda - 1)(\lambda - 9)$$

The singular values (in decreasing order) then are $\sigma_1 = \sqrt{9} = 3$, $\sigma_2 = \sqrt{1} = 1$ and $\Sigma = \begin{pmatrix} 3 & 0 \\ 0 & 1 \end{pmatrix}$.

The eigenvector belonging to the first eigenvalue of $\mathbf{M}^T\mathbf{M}$ follows from:

(continued)

Box 5.4 (continued)

$$\begin{pmatrix} 5 & -4 \\ -4 & 5 \end{pmatrix}\begin{pmatrix} x \\ y \end{pmatrix} = 9\begin{pmatrix} x \\ y \end{pmatrix} \Rightarrow \begin{pmatrix} 5x - 4y \\ -4x + 5y \end{pmatrix} = \begin{pmatrix} 9x \\ 9y \end{pmatrix}$$

$$\Rightarrow 5x - 4y = 9x \wedge -4x + 5y = 9y$$

$$\Rightarrow -4x - 4y = 0 \wedge -4x - 4y = 0$$

$$\Rightarrow x = -y$$

$$\Rightarrow \begin{pmatrix} x \\ y \end{pmatrix} = \begin{pmatrix} 1 \\ -1 \end{pmatrix}$$

To determine the first column of **V**, this eigenvector must be normalized (divided by its length; see Sect. 4.2.2.1): $\vec{v}_1 = \frac{1}{\sqrt{2}}\begin{pmatrix} 1 \\ -1 \end{pmatrix}$.

Similarly, the eigenvector belonging to the second eigenvalue of $M^T M$ can be derived to be $\vec{v}_2 = \frac{1}{\sqrt{2}}\begin{pmatrix} 1 \\ 1 \end{pmatrix}$, making $V = \frac{1}{\sqrt{2}}\begin{pmatrix} 1 & 1 \\ -1 & 1 \end{pmatrix}$. In this case, $\vec{v}_1$ and $\vec{v}_2$ are already orthogonal (see Sect. 4.2.2.1, making further adaptations to arrive at an orthonormal set of eigenvectors unnecessary. In case orthogonalization is necessary, Gram-Schmidt orthogonalization could be used (see Sect. 4.3.2).

In practice, to now determine U for this real 2×2 matrix **M**, it is most convenient to use that when $M = U\Sigma V^*(=U\Sigma V^T)$, $MV = U\Sigma$ or $MV\Sigma^{-1} = U$ (using that $V^*V = V^T V = I$ and Σ is real). Thus, the first column of U is equal to $\vec{u}_1 = \frac{1}{\sigma_1}M\vec{v}_1 = \frac{1}{3}\begin{pmatrix} 1 & -2 \\ 2 & -1 \end{pmatrix}\frac{1}{\sqrt{2}}\begin{pmatrix} 1 \\ -1 \end{pmatrix} = \frac{1}{\sqrt{2}}\begin{pmatrix} 1 \\ 1 \end{pmatrix}$ and the second column of U is equal to $\vec{u}_2 = \frac{1}{\sigma_2}M\vec{v}_2 = \frac{1}{1}\begin{pmatrix} 1 & -2 \\ 2 & -1 \end{pmatrix}\frac{1}{\sqrt{2}}\begin{pmatrix} 1 \\ 1 \end{pmatrix} = \frac{1}{\sqrt{2}}\begin{pmatrix} -1 \\ 1 \end{pmatrix}$, making $U = \frac{1}{\sqrt{2}}\begin{pmatrix} 1 & -1 \\ 1 & 1 \end{pmatrix}$. One can now verify that indeed $M = U\Sigma V^T$.

As promised, I will now discuss how SVD can be understood by virtue of this specific example. So let's see what V^T, Σ and U (in this order) do to some vectors on the unit circle: $\begin{pmatrix} 1 \\ 0 \end{pmatrix}$ in blue, $\frac{1}{\sqrt{2}}\begin{pmatrix} 1 \\ 1 \end{pmatrix}$ in yellow, $\begin{pmatrix} 0 \\ 1 \end{pmatrix}$ in red and $\frac{1}{\sqrt{2}}\begin{pmatrix} -1 \\ 1 \end{pmatrix}$ in green, as illustrated in Fig. 5.6.

V^T rotates these vectors over 45°. Σ subsequently scales the resulting vectors by factors of 3 in the x-direction and 1 in the y-direction, after which U performs another rotation over 45°. You can verify in Fig. 5.6 that the result of these three matrices applied successively is exactly the same as the result of applying **M** directly. Thus, given that **U** and **V** are rotation matrices and Σ is a scaling matrix, the intuitive understanding of SVD for real square matrices is that any such matrix transformation can be expressed as a rotation followed by a scaling followed by another rotation. It should be noted that for this intuitive understanding of SVD rotation should be understood to include *improper rotation*. Improper rotation combines proper rotation with reflection (an example is given in Exercise 5.16(a)).

Exercises

5.16. Calculate the SVD of

(a) $\begin{pmatrix} 2 & 1 \\ 1 & 2 \end{pmatrix}$

(b) $\begin{pmatrix} 2 & 0 \\ 0 & 1 \end{pmatrix}$

(continued)

5.17. For the SVD of **M** such that $M = U\Sigma V^*$ prove that the columns of **U** can be determined from the eigenvectors of **MM***.

Now that the mathematics of SVD and its intuitive meaning have been explained it is much easier to explain how SVD can be used to achieve data compression. When you've calculated the SVD of a matrix, representing e.g., the book reading behavior mentioned before, or the pixel values of a black-and-white image, you can compress the information by maintaining only the largest L singular values in Σ, setting all other singular values to zero, resulting in a sparser matrix Σ'. When you then calculate $M' = U\Sigma'V^*$, you'll find that only the first L columns of **U** and **V** remain relevant, as all other columns will be multiplied by zero. So instead of keeping the entire matrix **M** in memory, you only need to store a limited number of columns of **U** and **V**. The fraction of information that is maintained in this manner is determined by the fraction of the sum of the eigenvalues that are maintained divided by the sum of all eigenvalues. In the reference list at the end of this chapter you'll find links to an online SVD calculator and examples of data compression using SVD, allowing you to explore this application of complex matrix operations further.

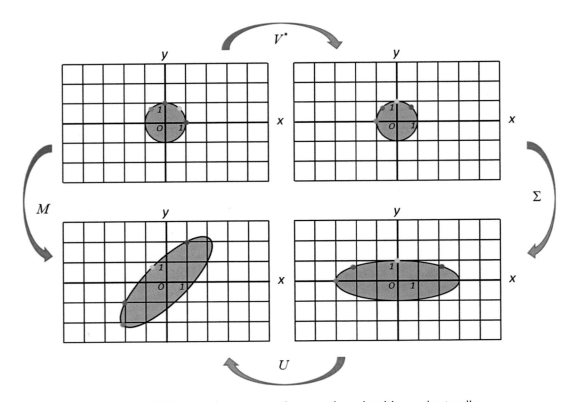

Fig. 5.6 Visualization of SVD for real square matrices to gain an intuitive understanding.

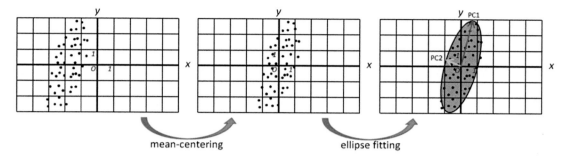

Fig. 5.7 Illustration of PCA in two dimensions. *Left*: example data distribution. *Middle*: mean-centered data distribution. *Right*: fitted ellipse and principal components PC1 (*red*) and PC2 (*yellow*).

Another method that is basically the same as SVD, that is also used for data compression but is formulated a bit differently is *principal component analysis* or PCA. PCA is typically described as a method that transforms data to a new coordinate system such that the largest variance is found along the first new coordinate (first principal component or PC), the next largest variance is found along the second new coordinate (second PC) etcetera. Let me explain this visually by an example in two dimensions. Suppose that the data is distributed as illustrated in Fig. 5.7 (left).

Here, we can order the data in a matrix $\mathbf{X}$ of dimensions $n \times 2$, where each row contains a data point (x_i, y_i), $i = 1 \ldots n$. In a p-dimensional space, our data matrix would have p columns. To perform PCA of the data (PC decomposition of this matrix), first the data is mean-centered (Fig. 5.7 (middle)), such that the mean in every dimension becomes zero. The next step of PCA can be thought of as fitting an ellipsoid (or ellipse in two dimensions) to the data (Fig. 5.7 (right)). The main axes of this ellipsoid represent the PCs; the longest axis the first PC, the next longest axis the second PC etcetera. Thus, long axes represent directions in the data with a lot of variance, short axes represent directions of little variance. The axes of the ellipsoid are represented by the orthonormalized eigenvectors of the *covariance matrix* of the data. The covariance matrix is proportional to $\mathbf{X}^{*}\mathbf{X}$ (or $\mathbf{X}^{T}\mathbf{X}$ for real matrices). The proportion of the variance explained by a PC is equal to the eigenvalue belonging to the corresponding eigenvector divided by the sum of all eigenvalues. This must sound familiar: compare it to the explanation of the information maintained during data compression using SVD in the previous paragraph. Mathematically, the PC decomposition of the $n \times p$ matrix $\mathbf{X}$ is given by $\mathbf{T} = \mathbf{X}\mathbf{W}$, where $\mathbf{W}$ (for 'weights') is a $p \times p$ matrix whose columns are the eigenvectors of $\mathbf{X}^{*}\mathbf{X}$ and $\mathbf{T}$ is an $n \times p$ matrix containing the component scores.

That PCA and SVD are basically the same can be understood as follows. We know now that the SVD of a matrix $\mathbf{M}$ is obtained by calculating the eigenvectors of $\mathbf{M}^{*}\mathbf{M}$ or $\mathbf{M}^{T}\mathbf{M}$. And indeed, when the SVD of $\mathbf{X}$ is given by $X = U\Sigma W^{T}$, then $X^{T}X = (U\Sigma W^{T})^{T}U\Sigma W^{T} = W\Sigma U^{T}U\Sigma W^{T} = W\Sigma^{2}W^{T}$. Thus, the eigenvectors of $\mathbf{X}^{T}\mathbf{X}$ are the columns of W, i.e. the right-singular vectors of $\mathbf{X}$. Or, in terms of the PC decomposition of $\mathbf{X}$: $T = XW = U\Sigma W^{T}W = U\Sigma$. Hence, each column of T is equal to the left singular vector of X times the corresponding singular value. Also notice the resemblance between Figs. 5.6 and 5.7.

As SVD, PCA is mostly used for data or dimensionality reduction (compression). By keeping only the first L components that explain most of the variance, high-dimensional data may become easier to store, visualize and understand. This is achieved by, again, truncating the matrices **T** and **W** to their first L columns. An example of dimensionality reduction from work in my own group (collaboration with Dr. O.E. Martinez Manzanera) is to use PCA to derive simple quantitative, objective descriptors of movement to aid clinicians in obtaining a diagnosis of specific movement disorders. In this work we recorded and analyzed fingertip movement of *ataxia* patients performing the so-called finger-to-nose test, in which they are asked to repeatedly move their index finger from their nose to the fingertip of an examiner. In healthy people, the fingertip describes a smooth curve during the finger-to-nose task, while in ataxia patients, the fingertip trajectory is much more irregular. To describe the extent of irregularity, we performed PCA on the coordinates of the fingertip trajectory, assuming that the trajectory of healthy participants would be mostly in a plane, implying that for them two PCs would explain almost all of the variance in the data. Hence, as an application of dimensionality reduction using PCA, the variance explained by the first two PCs provides a compact descriptor of the regularity of movement during this task. This descriptor, together with other movement features, was subsequently used in a classifier to distinguish patients with ataxia from patients with a milder coordination problem and from healthy people.

Finally, I would like to briefly introduce the method of *independent component analysis* or ICA, here. For many it is confusing what the difference is between PCA and ICA and when to use one or the other. As explained before, PCA minimizes the covariance (second order moment) of the data by rotating the basis vectors, thereby employing *Gaussianity* or normality of the data. This means that the data have to be normally distributed for PCA to work optimally. ICA, on the other hand, can determine independent components for non-Gaussian signals by minimizing higher order moments of the data (such as *skewness* and *kurtosis*) that describe how non-Gaussian their distribution is. Here, independence means that knowing the values of one component does not give any information about another component. Thus, if the data is not well characterized by its variance then ICA may work better than PCA. Or, vice versa, when the data are Gaussian, linear and *stationary*, PCA will probably work. In practice, when sensors measure signals of several sources at the same time, ICA is typically the method of choice. Examples of signals that are best analyzed by ICA are simultaneous sound signals (such as speech) that are picked up by several receivers (such as microphones) or electrical brain activity recorded by multiple *EEG* electrodes. The ICA process is also referred to as blind source separation. Actually, our brain does a very good job at separating sources, when we identify our own name being mentioned in another conversation amidst the buzz of a party (the so-called 'cocktail party effect'). There are many applications of ICA in science, such as identifying and removing noise signals from EEG measurements (see e.g. Islam et al. 2016) and identifying brain functional connectivity networks in functional MRI measurements (see e.g. Calhoun and Adali 2012). In the latter case PCA and ICA are actually sometimes used successively, when PCA is first used for dimensionality reduction followed by ICA for identifying connectivity networks.

Glossary

Adjacency matrix Matrix with binary entries (i,j) describing the presence (1) or absence (0) of a path between nodes i and j.

Adjoint Transpose of the cofactor matrix.

Airfoil Shape of an airplane wing, propeller blade or sail.

Ataxia A movement disorder or symptom involving loss of coordination.

Basis vector A set of (N-dimensional) basis vectors is linearly independent and any vector in N-dimensional space can be built as a linear combination of these basis vectors.

Boundary conditions Constraints for a solution to an equation on the boundary of its domain.

Cofactor matrix The (i,j)-element of this matrix is given by the determinant of the matrix that remains when the i-th row and j-th column are removed from the original matrix, multiplied by -1 if $i + j$ is odd.

Conjugate transpose Generalization of transpose; a transformation of a matrix $\mathbf{A}$ indicated by $\mathbf{A}^*$ with elements defined by $(A^*)_{ij} = \bar{a}_{ji}$.

Dense matrix A matrix whose elements are almost all non-zero.

Determinant Can be seen as a scaling factor when calculating the inverse of a matrix.

Diagonalization Decomposition of a matrix $\mathbf{M}$ such that it can be written as $M = VDV^{-1}$ where $\mathbf{V}$ is an invertible matrix and $\mathbf{D}$ is a diagonal matrix.

Diagonal matrix A matrix with only non-zero elements on the diagonal and zeroes elsewhere.

Discretize To represent an equation on a grid.

EEG Electroencephalography; a measurement of electrical brain activity.

Eigendecomposition To determine the eigenvalues and eigenvectors of a matrix.

Element As in 'matrix element': one of the entries in a matrix.

Gaussian Normally distributed.

Graph A collection of nodes or vertices with paths or edges between them whenever the nodes are related in some way.

Hadamard product Element-wise matrix product.

Identity matrix A square matrix with ones on the diagonal and zeroes elsewhere, often referred to as $\mathbf{I}$. The identity matrix is a special diagonal matrix.

Independent component analysis A method to determine independent components of non-Gaussian signals by minimizing higher order moments of the data.

Inverse The matrix $\mathbf{A}^{-1}$ such that $\mathbf{AA}^{-1} = \mathbf{A}^{-1}\mathbf{A} = \mathbf{I}$.

Invertable A matrix that has an inverse.

Kronecker product Generalization of the outer product (or tensor product or dyadic product) for vectors to matrices.

Kurtosis Fourth-order moment of data, describing how much of the data variance is in the tail of its distribution.

Laplace equation Partial differential equation describing the behavior of potential fields.

Left-singular vector Columns of $\mathbf{U}$ in the SVD of $\mathbf{M}$: $M = U\Sigma V^*$.

Leslie matrix Matrix with probabilities to transfer from one age class to the next in a population ecological model of population growth.

Logical matrix A matrix that only contains zeroes and ones (also: binary or Boolean matrix).

Matrix A rectangular array of (usually) numbers.

Network theory The study of graphs as representing relations between different entities, such as in a social network, brain network, gene network etcetera.

Order Size of a matrix.

Orthonormal Orthogonal vectors of length 1.

Partial differential equation An equation that contains functions of multiple variables and their partial derivatives (see also Chap. 6).

Principal component analysis Method that transforms data to a new coordinate system such that the largest variance is found along the first new coordinate (first PC), the then largest variance is found along the second new coordinate (second PC) etcetera.

Right-singular vector Columns of V in the SVD of $\mathbf{M}$: $M = U\Sigma V^{*}$

Root Here: a value of λ that makes the characteristic polynomial $|M - \lambda I|$ of the matrix $\mathbf{M}$ equal to zero.

Scalar function Function with scalar values.

Shearing To shift along one axis.

Singular value Diagonal elements of Σ in the SVD of $\mathbf{M}$: $M = U\Sigma V^{*}$.

Singular value decomposition The decomposition of an m × n rectangular matrix $\mathbf{M}$ into a product of three matrices such that $M = U\Sigma V^{*}$ where $\mathbf{U}$ is a unitary $m \times m$ matrix, Σ an $m \times n$ diagonal matrix with non-negative real entries and $\mathbf{V}$ another unitary $n \times n$ matrix.

Skewness Third-order moment of data, describing asymmetry of its distribution.

Skew-symmetric matrix A matrix $\mathbf{A}$ for which $a_{ij} = -a_{ji}$.

Sparse matrix A matrix with most of its elements equal to zero.

Stationary Time-dependent data for which the most important statistical properties (such as mean and variance) do not change over time.

Symmetric matrix A matrix $\mathbf{A}$ that is symmetric around the diagonal, i.e. for which $a_{ij} = a_{ji}$.

Transformation Here: linear transformation as represented by matrices. A function mapping a set onto itself (e.g. 2D space onto 2D space).

Transpose A transformation of a matrix $\mathbf{A}$ indicated by $\mathbf{A}^{\mathrm{T}}$ with elements defined by $(A^{T})_{ij} = a_{ji}$.

Triangular matrix A diagonal matrix extended with non-zero elements only above or only below the diagonal.

Unit matrix Identity matrix.

Symbols Used in This Chapter (in Order of Their Appearance)

$\mathbf{M}$ or M	Matrix (bold and capital letter in text, italic and capital letter in equations)
$(\cdot)_{ij}$	Element at position (i,j) in a matrix
$\sum_{k=1}^{n}$	Sum over k, from 1 to n
$\vec{}$	Vector
θ	Angle
$\circ$	Hadamard product, Schur product or pointwise matrix product
$\otimes$	Kronecker matrix product
$.^{T}$	(Matrix or vector) transpose
$.^{*}$	(Matrix or vector) conjugate transpose
$\dagger$	Used instead of * to indicate conjugate transpose in quantum mechanics
$.^{-1}$	(Matrix) inverse
$|\cdot|$	(Matrix) determinant

Overview of Equations, Rules and Theorems for Easy Reference

<u>Addition, subtraction and scalar multiplication of matrices</u>
Addition of matrices **A** and **B** (of the same size):

$$(A + B)_{ij} = a_{ij} + b_{ij}$$

Subtraction of matrices **A** and **B** (of the same size):

$$(A - B)_{ij} = a_{ij} - b_{ij}$$

Multiplication of a matrix **A** by a scalar s:

$$(sA)_{ij} = sa_{ij}$$

<u>Basis vector principle</u>
Any vector $\begin{pmatrix} a \\ b \end{pmatrix}$ (in 2D space) can be built from the basis vectors $\begin{pmatrix} 1 \\ 0 \end{pmatrix}$ and $\begin{pmatrix} 0 \\ 1 \end{pmatrix}$ by a linear combination as follows: $\begin{pmatrix} a \\ b \end{pmatrix} = a \begin{pmatrix} 1 \\ 0 \end{pmatrix} + b \begin{pmatrix} 0 \\ 1 \end{pmatrix}$.

The same principle holds for vectors in higher dimensions.

<u>Rotation matrix (2D)</u>
The transformation matrix that rotates a vector around the origin (in 2D) over an angle θ (counter clockwise) is given by $\begin{pmatrix} \cos\theta & -\sin\theta \\ \sin\theta & \cos\theta \end{pmatrix}$.

<u>Shearing matrix (2D)</u>
$\begin{pmatrix} 1 & k \\ 0 & 1 \end{pmatrix}$: shearing along the x-axis (y-coordinate remains unchanged)
$\begin{pmatrix} 1 & 0 \\ k & 1 \end{pmatrix}$: shearing along the y-axis (x-coordinate remains unchanged)

<u>Matrix product</u>
Multiplication **AB** of an $m \times n$ matrix **A** with an $n \times p$ matrix **B**:

$$(AB)_{ij} = \sum_{k=1}^{n} a_{ik}b_{kj}$$

Hadamard product, Schur product or pointwise product:

$$(A \circ B)_{ij} = a_{ij}b_{ij}$$

Kronecker product:

$$A \otimes B = \begin{pmatrix} a_{11}B & a_{12}B & \cdots & a_{1n}B \\ a_{21}B & a_{22}B & \cdots & a_{2n}B \\ \vdots & \vdots & \ddots & \vdots \\ a_{m1}B & a_{m2}B & \cdots & a_{mn}B \end{pmatrix}$$

Special matrices

Hermitian matrix: $\mathbf{A} = \mathbf{A}^*$

normal matrix: $\mathbf{A}^*\mathbf{A} = \mathbf{A}\mathbf{A}^*$

unitary matrix: $\mathbf{A}\mathbf{A}^* = \mathbf{I}$

where $(A^*)_{ij} = \bar{a}_{ji}$ defines the conjugate transpose of $\mathbf{A}$.

Matrix inverse

For a square matrix $\mathbf{A}$ the inverse $A^{-1} = \frac{1}{\det(A)} adj(A)$, where $\det(\mathbf{A})$ is the determinant of $\mathbf{A}$ and $adj(\mathbf{A})$ is the adjoint of $\mathbf{A}$ (see Sect. 5.3.1).

Eigendecomposition

An eigenvector $\vec{v}$ of a square matrix $\mathbf{M}$ is determined by:

$$M\vec{v} = \lambda\vec{v},$$

where λ is a scalar known as the eigenvalue

Diagonalization

Decomposition of a square matrix $\mathbf{M}$ such that:

$$M = VDV^{-1}$$

where $\mathbf{V}$ is an invertible matrix and $\mathbf{D}$ is a diagonal matrix

Singular value decomposition

Decomposition of an m $\times$ n rectangular matrix $\mathbf{M}$ such that:

$$M = U\Sigma V^*$$

where $\mathbf{U}$ is a unitary $m \times m$ matrix, $\mathbf{\Sigma}$ an $m \times n$ diagonal matrix with non-negative real entries and $\mathbf{V}$ another unitary $n \times n$ matrix.

Answers to Exercises

5.1. (a)

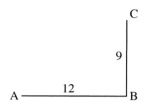

(b) The direct distance between cities A and C can be calculated according to Pythagoras' theorem as $\sqrt{12^2 + 9^2} = \sqrt{144 + 81} = \sqrt{225} = 15$. Hence, the distance matrix becomes $\begin{pmatrix} 0 & 12 & 15 \\ 12 & 0 & 9 \\ 15 & 9 & 0 \end{pmatrix}$.

(c)

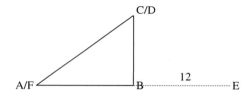

5.2. The sum and difference of the pairs of matrices are:

(a) $\begin{pmatrix} 5 & 2 \\ 2 & 15 \end{pmatrix}$ and $\begin{pmatrix} 1 & 6 \\ -4 & 1 \end{pmatrix}$

(b) $\begin{pmatrix} 7 & -4 & 6 \\ -1 & 4 & 1 \\ -4 & 6 & 2 \end{pmatrix}$ and $\begin{pmatrix} -1 & -10 & 2 \\ -3 & 8 & 9 \\ 6 & -10 & -20 \end{pmatrix}$

(c) $\begin{pmatrix} 2 & 1.6 & -1 \\ 5.1 & 1 & -2 \end{pmatrix}$ and $\begin{pmatrix} 0.4 & 4.8 & -2 \\ 1.7 & 3.6 & -4.4 \end{pmatrix}$

5.3. (a) $\begin{pmatrix} 6 & 0 \\ 7 & -4 \\ 8 & 10 \end{pmatrix}$

(b) $\begin{pmatrix} 3.5 & 0 \\ 2.9 & 0.7 \end{pmatrix}$

5.4. Possibilities for multiplication are **AB**, **AC**, **BD**, **CB** and **DA**.

5.5. (a) 2×7
    (b) 2×1
    (c) 1×1

5.6. (a) $\mathbf{AB} = \begin{pmatrix} 18 & 22 \\ 22 & 58 \end{pmatrix}$, $\mathbf{BA} = \begin{pmatrix} 8 & -8 \\ 2 & 68 \end{pmatrix}$

(b) $\mathbf{BA} = \begin{pmatrix} 8 & -14 \\ 3 & -11 \\ -20 & 61 \end{pmatrix}$

(c) no matrix product possible

5.7. (a) $\begin{pmatrix} 2 & 6 \\ -4 & 1 \end{pmatrix}$

(b) $\begin{pmatrix} 2.8 & -2.7 & 0.5 \\ 12 & -0.7 & -4 \end{pmatrix}$

5.8. (a) $\begin{pmatrix} 3 & -1 & 4 & 6 & -2 & 8 \\ -3 & 1 & -4 & 3 & -1 & 4 \end{pmatrix}$

(b) $\begin{pmatrix} 0 & -2 & -4 & 0 & -3 & -6 \\ -6 & -8 & -10 & -9 & -12 & -15 \\ -12 & -14 & -16 & -18 & -21 & -24 \end{pmatrix}$

5.9. (a) symmetric, logical
(b) sparse, upper-triangular
(c) skew-symmetric
(d) upper-triangular
(e) diagonal, sparse
(f) identity, diagonal, logical, sparse

5.10. (a) $\begin{pmatrix} 1 & i & 5 \\ 2 & 1 & 4 - 5i \\ 3 & -3 + 2i & 3 \end{pmatrix}$

(b) $\begin{pmatrix} 1 & -1 & 5 \\ 2 & 1 & 4 \\ 3 & -3 & 0 \end{pmatrix}$

(c) $\begin{pmatrix} 4 & 19 - i & 8i \\ 0 & -3 & -11 + i \\ 3 + 2i & -3 & 17 \end{pmatrix}$

5.11. Using Cramer's rule we find that $x = \dfrac{D_x}{D} = \begin{vmatrix} c & b \\ f & e \end{vmatrix} / \begin{vmatrix} a & b \\ d & e \end{vmatrix} = \dfrac{ce - bf}{ae - bd}$ and

$y = \dfrac{D_y}{D} = \begin{vmatrix} a & c \\ d & f \end{vmatrix} / \begin{vmatrix} a & b \\ d & e \end{vmatrix} = \dfrac{af - cd}{ae - bd}$. From Sect. 5.3.1 we obtain that the inverse of the

matrix $\begin{pmatrix} a & b \\ d & e \end{pmatrix}$ is equal to $\dfrac{1}{ae - bd} \begin{pmatrix} e & -b \\ -d & a \end{pmatrix}$ and the solution to the system of linear equations is $\dfrac{1}{ae-bd}\begin{pmatrix} e & -b \\ -d & a \end{pmatrix}\begin{pmatrix} c \\ f \end{pmatrix} = \begin{pmatrix} (ce-bf)/(ae-bd) \\ (af-cd)/(ae-bd) \end{pmatrix}$ which is the same as the solution obtained using Cramer's rule.

5.12. (a) $x = -11, y = -5\frac{3}{5}$

(b) $D = \begin{vmatrix} 4 & -2 & -2 \\ 2 & 8 & 4 \\ 30 & 12 & -4 \end{vmatrix} = 4(8 \cdot -4 - 4 \cdot 12) - \cdot -2(2 \cdot -4 - 30 \cdot 4) + \cdot -2(2 \cdot 12$

$-30 \cdot 8) = -144$

$$D_x = \begin{vmatrix} 10 & -2 & -2 \\ 32 & 8 & 4 \\ 24 & 12 & -4 \end{vmatrix} = -1632 \quad D_y = \begin{vmatrix} 4 & 10 & -2 \\ 2 & 32 & 4 \\ 30 & 24 & -4 \end{vmatrix} = 2208$$

$$D_z = \begin{vmatrix} 4 & -2 & 10 \\ 2 & 8 & 32 \\ 30 & 12 & 24 \end{vmatrix} = -4752$$

Thus, $x = \dfrac{D_x}{D} = \dfrac{-1632}{-144} = 11\frac{1}{3}$, $\qquad y = \dfrac{D_y}{D} = \dfrac{2208}{-144} = -15\frac{1}{3}$ and

$z = \dfrac{D_z}{D} = \dfrac{-4752}{-144} = 33$

5.13. (a) $x = 4, y = 0$

(b) $x = 2, y = -1, z = 1$

5.14. This shearing matrix shears along the x-axis and leaves y-coordinates unchanged (see Sect. 5.2.2). Hence, all vectors along the x-axis remain unchanged due to this transformation. The eigenvector is thus $\begin{pmatrix} 1 \\ 0 \end{pmatrix}$ with eigenvalue 1 (since the length of the eigenvector is unchanged due to the transformation).

5.15. (a) $\lambda_1 = 7$ with eigenvector $\begin{pmatrix} 1 \\ 0 \\ 0 \end{pmatrix}$ (x = x, y = 0, z = 0), $\lambda_2 = -19$ with eigenvector

$\begin{pmatrix} 0 \\ 1 \\ 0 \end{pmatrix}$ (x = 0, y = y, z = 0) and $\lambda_3 = 2$ with eigenvector $\begin{pmatrix} 0 \\ 0 \\ 1 \end{pmatrix}$ (x = 0, y = 0, z = z).

(b) $\lambda = 3$ (double) with eigenvector $\begin{pmatrix} 1 \\ 1 \end{pmatrix}$ (y = x).

(c) $\lambda_1 = 1$ with eigenvector $\begin{pmatrix} -1 \\ 1 \\ 1 \end{pmatrix}$ (x = -z, y = z), $\lambda_2 = -1$ with eigenvector $\begin{pmatrix} 1 \\ 1 \\ 5 \end{pmatrix}$

(5x = z, 5y = z) and $\lambda_3 = 3$ with eigenvector $\begin{pmatrix} 1 \\ 1 \\ 1 \end{pmatrix}$ (x = y = z).

(d) $\lambda_1 = 3$ with eigenvector $\begin{pmatrix} 1 \\ -2 \end{pmatrix}$ ($y = -2x$) and $\lambda_2 = 7$ with eigenvector $\begin{pmatrix} 1 \\ 2 \end{pmatrix}$ ($y = 2x$).

5.16. (a) To determine the SVD, we first determine the eigenvalues and eigenvectors of $\mathbf{M}^T \mathbf{M}$ to get the singular values and right-singular vectors of $\mathbf{M}$. Thus, we determine its characteristic equation as the determinant

$$\left| \begin{pmatrix} 2 & 1 \\ 1 & 2 \end{pmatrix} \begin{pmatrix} 2 & 1 \\ 1 & 2 \end{pmatrix} - \lambda \begin{pmatrix} 1 & 0 \\ 0 & 1 \end{pmatrix} \right| = \left| \begin{pmatrix} 5 & 4 \\ 4 & 5 \end{pmatrix} - \lambda \begin{pmatrix} 1 & 0 \\ 0 & 1 \end{pmatrix} \right| =$$

$$\left| \begin{pmatrix} 5 - \lambda & 4 \\ 4 & 5 - \lambda \end{pmatrix} \right| = (5 - \lambda)^2 - 16 = \lambda^2 - 10\lambda + 9 = (\lambda - 1)(\lambda - 9)$$

Thus, the singular values are $\sigma_1 = \sqrt{9} = 3$ and $\sigma_1 = \sqrt{1} = 1$ and $\Sigma = \begin{pmatrix} 3 & 0 \\ 0 & 1 \end{pmatrix}$.

The eigenvector belonging to the first eigenvalue of $\mathbf{M}^T \mathbf{M}$ follows from:

$$\begin{pmatrix} 5 & 4 \\ 4 & 5 \end{pmatrix} \begin{pmatrix} x \\ y \end{pmatrix} = 9 \begin{pmatrix} x \\ y \end{pmatrix} \Rightarrow \begin{pmatrix} 5x + 4y \\ 4x + 5y \end{pmatrix} = \begin{pmatrix} 9x \\ 9y \end{pmatrix}$$

$$\Rightarrow 5x + 4y = 9x \wedge 4x + 5y = 9y$$

$$\Rightarrow -4x + 4y = 0 \wedge 4x - 4y = 0$$

$$\Rightarrow x = y$$

$$\Rightarrow \begin{pmatrix} x \\ y \end{pmatrix} = \begin{pmatrix} 1 \\ 1 \end{pmatrix}$$

To determine the first column of $\mathbf{V}$, this eigenvector must be normalized (divided by its length; see Sect. 4.2.2.1) and thus $\vec{v}_1 = \frac{1}{\sqrt{2}} \begin{pmatrix} 1 \\ 1 \end{pmatrix}$.

Similarly, the eigenvector belonging to the second eigenvalue of $\mathbf{M}^T \mathbf{M}$ can be derived to be $\vec{v}_2 = \frac{1}{\sqrt{2}} \begin{pmatrix} 1 \\ -1 \end{pmatrix}$, making $V = \frac{1}{\sqrt{2}} \begin{pmatrix} 1 & 1 \\ 1 & -1 \end{pmatrix}$. In this case, $\vec{v}_1$ and $\vec{v}_2$ are already orthogonal (see Sect. 4.2.2.1), making further adaptations to arrive at an orthonormal set of eigenvectors unnecessary.

To determine U we use that $\vec{u}_1 = \frac{1}{\sigma_1} M \vec{v}_1 = \frac{1}{3} \begin{pmatrix} 2 & 1 \\ 1 & 2 \end{pmatrix} \frac{1}{\sqrt{2}} \begin{pmatrix} 1 \\ 1 \end{pmatrix} = \frac{1}{\sqrt{2}} \begin{pmatrix} 1 \\ 1 \end{pmatrix}$ and

$\vec{u}_2 = \frac{1}{\sigma_2} M \vec{v}_2 = \frac{1}{1} \begin{pmatrix} 2 & 1 \\ 1 & 2 \end{pmatrix} \frac{1}{\sqrt{2}} \begin{pmatrix} 1 \\ -1 \end{pmatrix} = \frac{1}{\sqrt{2}} \begin{pmatrix} 1 \\ -1 \end{pmatrix}$, making $U = \frac{1}{\sqrt{2}} \begin{pmatrix} 1 & 1 \\ 1 & -1 \end{pmatrix}$.

You can now verify yourself that indeed $M = U \Sigma V^*$.

(b) Taking a similar approach, we find that

$$U = \begin{pmatrix} 1 & 0 \\ 0 & 1 \end{pmatrix}, \Sigma = \begin{pmatrix} 2 & 0 \\ 0 & 1 \end{pmatrix} \text{ and } V = \begin{pmatrix} 1 & 0 \\ 0 & 1 \end{pmatrix}.$$

5.17. If $M = U\Sigma V^*$ and using that both **U** and **V** are unitary (see Table 5.1), then $MM^* = U\Sigma V^*(U\Sigma V^*)^* = U\Sigma V^* V\Sigma U^* = U\Sigma^2 U^*$. Right-multiplying both sides of this equation with U and then using that Σ^2 is diagonal, yields $MM^* U = U\Sigma^2 = \Sigma^2 U$. Hence, the columns of U are eigenvectors of **MM*** (with eigenvalues equal to the diagonal elements of Σ^2).

References

Online Sources of Information: History

http://www-history.mcs.st-and.ac.uk/history/HistTopics/Matrices_and_determinants.html#86

Online Sources of Information: Methods

https://en.wikipedia.org/wiki/Matrix_(mathematics)
https://en.wikipedia.org/wiki/Eigenvalues_and_eigenvectors
https://en.wikipedia.org/wiki/Singular_value_decomposition
https://en.wikipedia.org/wiki/Rotation_matrix
https://www.cs.helsinki.fi/u/ahyvarin/whatisica.shtml
http://comnuan.com/cmnn01004/ (one of many online matrix calculators)
http://andrew.gibiansky.com/blog/mathematics/cool-linear-algebra-singular-value-decomposition/
    (examples of information compression using SVD)

Books

R.A. Horn, C.R. Johnson, *Topics in Matrix Analysis* (Cambridge University Press, Cambridge, 1994)

W.H. Press, B.P. Flannery, S.A. Teukolsky, W.T. Vetterling, *Numerical Recipes, the Art of Scientific Computing. (Multiple Versions for Different Programming Languages)* (Cambridge University Press, Cambridge)

Y. Saad. Iterative Methods for Sparse Linear Systems, SIAM, New Delhi 2003. http://www-users.cs.umn.edu/~saad/IterMethBook_2ndEd.pdf

Papers

V.D. Calhoun, T. Adali, Multisubject independent component analysis of fMRI: a decade of intrinsic networks, default mode, and neurodiagnostic discovery. IEEE Rev. Biomed. Eng. **5**, 60–73 (2012)

M.K. Islam, A. Rastegarnia, Z. Yang, Methods for artifact detection and removal from scalp EEG: a review. Neurophysiol. Clin. **46**, 287–305 (2016)

6

Limits and Derivatives

Branislava Ćurčić-Blake

After reading this chapter you know:

- why you need limits and derivatives,
- what derivatives and limits are,
- how to determine a limit,
- how to calculate a derivative,
- what the maximum and minimum of a function are in terms of derivatives and
- what the slope of a function is.

6.1 Introduction to Limits

Whenever we are dealing with an optimization problem, such as finding the brightest pixel in a picture, the strongest activation in an fMRI statistical map, the minimum of the fitting error or the maximum likelihood we need to calculate a *derivative*. A derivative provides a measure of the rate of change of a function and to find the minimum or maximum of a function we need to know where the rate of change of the function is zero. These concepts are illustrated in Fig. 6.1.

Thus, to find *extrema* of a function we determine the *roots* of the derivative of the function. To understand how derivatives are defined we need to understand *limits* of functions first. A limit is a value that a function approaches when an input variable (or *argument*) of that function approaches a value. This sounds rather abstract, so what does it mean? The best way to clarify this is by an example.

B. Ćurčić-Blake (✉)
Neuroimaging Center, University Medical Center Groningen, Groningen, The Netherlands
e-mail: b.curcic@umcg.nl

© Springer International Publishing AG 2017   **163**
N. Maurits, B. Ćurčić-Blake, *Math for Scientists*, DOI 10.1007/978-3-319-57354-0_6

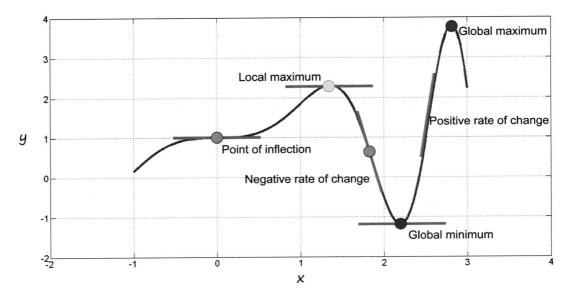

Fig. 6.1 Illustration of how the minima and maxima of a function are related to the rate of change of a function. Here the function $f(x)=x\sin(x^2)+1$ is plotted for $x\in[-1,3]$. It has a global maximum and minimum (*red dots*), a local maximum (*yellow dot*) and *points of inflection* (points where the curvature changes sign; two are indicated by *blue dots*). The slope of the tangent lines (*red*), which is the local rate of change of the function, is equal to the derivative of the function. It can be seen that it can be positive, negative or zero. At maxima, minima, and at some (but not all) points of inflection it is zero.

Example 6.1

What is the limit of the function $f(x)=x+3.3$ when x approaches 3?

We can examine this by looking at the value of the function when x is close to 3. Let's consider x equal to 2.5, 2.9, 2.99, 2.999. . .. In the table below, the values of x are in the left column, and the associated values of the function are in the right column.

x	$y=x+3.3$
2.5	5.8
2.9	6.2
2.99	6.29
2.999	6.299
2.9999	6.2999

Thus, the function approaches 6.3 if its variable x approaches 3. However, we only considered the function for x approaching 3 from below (from the left). Since the direction of approach was NOT specified, we also need to consider the approach from above (from the right) e.g. for values of x equal to 3.5, 3.1, 3.01, 3.001, 3.0001 etc. Also, when approaching $x = 3$ from above, the value for this function is 6.3:

x	$y=x+3.3$
3.5	6.8
3.1	6.4
3.01	6.31
3.001	6.301
3.0001	6.3001

This example illustrates what it means to determine the limit of a function. Importantly, unless otherwise specified, to determine the limit of a function, one needs to determine the value of the function both for the variable approaching a given value from above (from the right or from higher values) as well as from below (from the left or from lower values). For this first example, we can also calculate the limit algebraically by substituting $x = 3$ to find that $f(x) = 6.3$. However, this is not always the case, as the next example illustrates.

Example 6.2

What is the limit of $f(x) = \frac{x^2-4}{x-2}$ when x approaches 2?

Note that we cannot calculate the limit algebraically straightaway, as the function is not defined for $x = 2$. The nominator is 0 for this value of x and division by zero is not defined. We will discuss this in more detail in Sect. 6.6 about continuity. To determine this limit we make tables as in Example 6.1 with x approaching 2 from below and from above:

x	$y = \dfrac{x^2 - 4}{x - 2}$
1.5	3.5
1.9	3.9
1.99	3.99
1.999	3.999
1.9999	3.9999

x	$y = \dfrac{x^2 - 4}{x - 2}$
2.5	4.5
2.1	4.1
2.01	4.01
2.001	4.001
2.0001	4.0001

Thus, since $f(x)$ approaches 4 for x approaching 2 from below as well as from above, the limit of this function is 4. The graph of this function is given in Fig. 6.2. In this case the limit can also be calculated algebraically by first manipulating the function. As the numerator is divisible by $x - 2$, the function can be simplified to $f(x)=x+2$, which is the line in Fig. 6.2. Substitution of $x = 2$ then yields $(x)=4$.

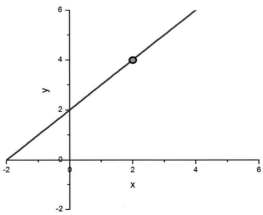

Fig. 6.2 The *circle* indicates the limit of this function when x approaches 2. This is a *removable discontinuity* (see Sect. 6.6).

6.2   Intuitive Definition of Limit

Mathematically, the limit is expressed as

$$\lim_{x \to a} f(x) = L$$

which can be read as 'the limit of f(x) when x approaches a is L'. Intuitively, this means that f(x) can become arbitrarily close to L, when bringing x sufficiently close to a. For the limit to exist, it has to be the same for x approaching a from below (the left) as well as from above (the right). Note that f(x) itself does not need to be defined at $x = a$, as was illustrated in Example 6.2.

In some cases, the limit of a function may only exist from one side (from above or from below) or it may exist from both sides but have different values. For these cases it is important to have a separate notation of *one-sided limits*. **The limit from the right** is the value L that a function f(x) approaches when x approaches a **from higher values** or from the right and is indicated by:

$$\lim_{x \to a^+} f(x) = L \quad \text{or} \quad \lim_{x \downarrow a} f(x) = L$$

Similarly, **the limit from the left** is the value L that a function f(x) approaches when x approaches a **from lower values** or from the left and is indicated by:

$$\lim_{x \to a^-} f(x) = L \quad \text{or} \quad \lim_{x \uparrow a} f(x) = L$$

The next example provides a case where the limits from the right and from the left differ.

Example 6.3

Determine $\lim_{x \to 2} \frac{x^3}{x-2}$.

We can determine whether the limit exists by making tables for x approaching 2 from the left and from the right:

x	$y = \dfrac{x^3}{x-2}$
1.5	−6.75
1.9	−68.59
1.99	−788.059
1.999	−7988.05

x	$y = \dfrac{x^3}{x-2}$
2.5	31.25
2.1	92.61
2.01	812.06
2.001	8012.006

(continued)

Example 6.3 (continued)

As you can see, the limits from the left and from the right exist, but they are not the same. The limit from the left keeps increasing negatively and is approaching $-\infty$ (minus *infinity*), whereas the limit from the right keeps increasing positively and is approaching $+\infty$. Thus, the function has no limit for x approaching 2.

Exercise

6.1. Determine the following limits using tables to approach the limit from the left and from the right

(a) $\lim_{x \to 7} (3x - 3)$

(b) $\lim_{x \to 7} \sqrt{3x^2 - 3}$

(c) $\lim_{x \to 7} \frac{3x-3}{x+5}$

(d) $\lim_{x \to 3} f(x)$ when $f(x) = \begin{cases} x - 2, & x \le 3 \\ 3x, & x > 3 \end{cases}$

6.3 Determining Limits Graphically

In certain cases, such as when functions are different for different domains, it may be easier to determine (the existence of) limits graphically. Functions that are different for (two) different domains are functions of the type:

$$f(x) = \begin{cases} f_1(x), \text{ for } x \in (-\infty, a] \\ f_2(x), \text{ for } x \in (a, \infty) \end{cases}$$

where a is the value of x defining the boundary between the two domains in this case. Domains can be defined in different ways (see Table 2.3 for more examples), and the number of domains may vary. For example, it is also possible to have a function with three domains, such as:

$$f(x) = \begin{cases} f_1(x), & for x \in (-\infty, a] \\ f_2(x), & for x \in (a, b] \\ f_3(x), & for x \in (b, \infty) \end{cases}$$

For these functions, it helps to plot their graph to determine whether the limit exists, i.e. to determine whether the limits from the left and the right are the same. We will illustrate this with two examples:

Example 6.4

Determine $\lim\limits_{x \to 1} f(x)$ for the function:

$$f(x) = \begin{cases} x - 2, & \textit{for } x \in (-\infty, 1] \\ x^2, & \textit{for } x \in (1, \infty) \end{cases}$$

When plotting the graph of this function,

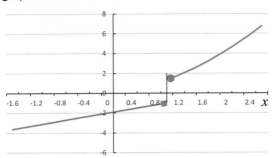

we observe that the limits from the left and from the right are different. Thus, this function has no limit for x approaching 1.

Example 6.5

Determine $\lim\limits_{x \to 2} f(x)$ for the function:

$$f(x) = \begin{cases} x + 3, & \textit{for } x \in (-\infty, 2] \\ -x + 7, & \textit{for } x \in (2, \infty) \end{cases}$$

Again, we plot the graph of this function:

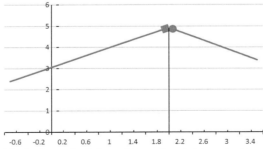

Here we see that the limits from the left and from the right are both equal to 5. Thus, the limit of this function for x approaching 2 is 5.

For these two examples we showed how plotting the graph of a function can help determine the existence of a limit. However, also for functions that are different for different domains, their limits can often be obtained algebraically. This is the case for Example 6.5, but also for the function in the next example that is defined on three domains.

Example 6.6

Determine the limit for x approaching a) ½; b) 1; c) 1.001; d) 3.5, for the following function

$$f(x) \begin{cases} 2x, & -2 < x < 1 \\ 1, & x = 1 \\ 2x, & 1 < x < 3 \end{cases}$$

To determine the limits, the relevant domain has to be identified. Once that is done, the limits can be determined algebraically as follows:

a) $x = \frac{1}{2}$ is in the domain $(-2, 1)$ where $f(x) = 2x$. Thus $\lim_{x \to \frac{1}{2}} f(x) = 2 \cdot \frac{1}{2} = 1$.

b) We have to determine the limit for x approaching 1 from the left and for x approaching 1 from the right. In both cases, $f(x) = 2x$, as the function is only 1 for $x = 1$. Thus $\lim_{x \to 1} f(x) = 2 \cdot 1 = 2$.

c) $x = 1.001$ is in the domain $(1, 3)$ where $f(x) = 2x$. Thus $\lim_{x \to 1.001} f(x) = 2 \cdot 1.001 = 2.002$.

d) The function is not defined for $x = 3.5$. Thus $\lim_{x \to 3.5} f(x)$ does not exist.

6.4   Arithmetic Rules for Limits

In this section we provide the basic arithmetic rules for calculating limits. They are provided for two-sided limits, but are similar for one-sided limits.

Box 6.1 Arithmetic rules for limits

1. $\lim_{x \to c} a \cdot f(x) = a \cdot \lim_{x \to c} f(x)$, when a is constant

2. $\lim_{x \to c} [f(x) \pm g(x)] = \lim_{x \to c} f(x) \pm \lim_{x \to c} g(x)$

3. $\lim_{x \to c} [f(x)g(x)] = \lim_{x \to c} f(x) \cdot \lim_{x \to c} g(x)$

4. $\lim_{x \to c} \left[\frac{f(x)}{g(x)} \right] = \frac{\lim_{x \to c} f(x)}{\lim_{x \to c} g(x)}$, if and only if $\lim_{x \to c} g(x) \neq 0$

5. $\lim_{x \to c} f(x)^n = \left[\lim_{x \to c} f(x) \right]^n$, $n \in \mathbb{R}$

6. $\lim_{x \to c} a = a$, when a is constant

7. $\lim_{x \to c} x = c$

Example 6.7

Determine $\lim_{x \to 5} \frac{x^2 - 4x - 5}{x - 4}$.

Applying the arithmetic rules in Box 6.1 we can rewrite the limit as follows:

$$\lim_{x \to 5} \frac{x^2 - 4x - 5}{x - 4} = \frac{\lim_{x \to 5} (x^2 - 4x - 5)}{\lim_{x \to 5} (x - 4)} = \frac{0}{1} = 0$$

Exercise

6.2. Determine the following limits using the arithmetic rules in Box 6.1

    a) $\lim\limits_{x \to -2} \frac{x^2+6x+7}{x+3}$

    b) $\lim\limits_{x \to -2} \frac{x^2-9}{x+3}$

    c) $\lim\limits_{x \to 4} \frac{x^2-1}{x^2+x-2}$

    d) $\lim\limits_{x \to 1} \frac{x^2-6x+9}{x^2-9}$

6.5 Limits at Infinity

We already encountered infinity in the context of limits in Example 6.3, where we calculated $\lim\limits_{x \to 2} \frac{x^3}{x-2}$. We found that the limits from the left and right were different and kept increasing, the closer we got to $x = 2$. They were approaching $-\infty$ and ∞, from the left and from the right, respectively. In the present context infinity, or ∞, is an abstract concept that can be thought of as a number larger than any number. Yet, even though ∞ is an abstract concept, we can work with it and actually, do mathematics with it, as we will see later in this section.

Example 6.8

Determine $\lim\limits_{x \to 0} \frac{1}{x}$.

    The function $\frac{1}{x}$ is not defined for $x = 0$ (division by zero). Hence, we first plot the graph of this function to understand what's happening (Fig. 6.3):

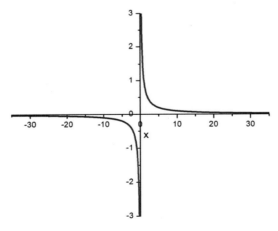

Fig. 6.3 Graph of the function $f(x) = \frac{1}{x}$. This is an example of an *asymptotic discontinuity* (see Sect. 6.6).

(continued)

Example 6.8 (continued)

We can observe that the limit from the right approaches ∞:

$$\lim_{x \to 0^+} \frac{1}{x} = \infty$$

whereas the limit from the left approaches $-\infty$:

$$\lim_{x \to 0^-} \frac{1}{x} = -\infty$$

Thus, the limit has different values when x approaches zero from the left or from the right and it does not exist.

We can also determine limits of functions when their argument approaches ∞: $\lim_{x \to \infty}$. In such cases we have to examine what happens with the function when its input variable becomes larger and larger. In doing so, we have to consider how fast parts of the function approach ∞, as illustrated in the next examples.

Example 6.9

Determine $\lim_{x \to \infty} (x^2 - 3x)$.

We can decompose this limit according to the arithmetic rules for limits in Box 6.1:

$$\lim_{x \to \infty} (x^2 - 3x) = \lim_{x \to \infty} x^2 - \lim_{x \to \infty} (3x)$$

Now both limits approach ∞ and thus it seems that $\lim_{x \to \infty} (x^2 - 3x) = \infty - \infty = 0$. However, there are different 'degrees' of infinity which can be understood as follows. For $x > 3$, $x^2 > 3x$, thus we can say that x^2 approaches infinity faster than $3x$. This is why the limit in this example is $+\infty$:

$$\lim_{x \to \infty} (x^2 - 3x) = \infty$$

Example 6.10

Determine $\lim_{x \to \infty} \frac{(x^2 - 3x)}{(3x^2 + 3x)}$.

The rule of thumb here is to identify the highest power of the argument. In this rational polynomial function the highest power of x is x^2. The highest power will determine the outcome of the limit as it results in the fastest approach of infinity and will dominate all lower powers. Thus, all elements with lower power can be neglected and we only need to consider the elements of quadratic power in this case:

$$\lim_{x \to \infty} \frac{(x^2 - 3x)}{(3x^2 + 3x)} = \lim_{x \to \infty} \frac{x^2}{3x^2} = \frac{1}{3}$$

6.6 Application of Limits: Continuity

An application of limits in mathematics itself is in the definition of continuity of a function. Intuitively, a function is *continuous* on a given domain if its graph is fully connected and has no jumps, gaps of holes, or slightly more formally, if for small enough changes in its arguments, the change in value of the function is arbitrarily small. One definition of continuity of a function $f(x)$ in a point c is:

$$\lim_{x \to c} f(x) = f(c)$$

Given what we now know about limits this definition implies three things:

1) f is defined at c
2) the limit exists
3) the value of this limit is equal to $f(c)$

This implies that $f(x)$ only has to be continuous for all arguments in its domain for it to be continuous. As an example $f(x) = \frac{1}{x}$ is a continuous function, even though its limit at $x = 0$ does not exist (Example 6.7). It is still continuous because zero does not belong to its *domain* (the set of arguments for which the function is defined). However, it has an (*asymptotic*) discontinuity at $x = 0$, which will be discussed later in this section.

The concept of continuity is very important, as it provides a basis for other mathematical concepts, such as for example the derivatives that are introduced later in this chapter. In practice, it for example helps understanding the difference between digital and analogue recordings. Often—but not always—analogue recordings are continuous, while digital recordings are always discontinuous or *discrete* as they *sample* a data stream at a certain sampling rate and not at every point in time.

For most mathematical applications, we deal with functions that are discontinuous at certain points only. Several types of such discontinuities exist and we actually already encountered a few:

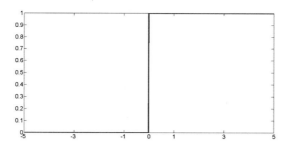

Fig. 6.4 Heaviside step function.

1. <u>Point or removable discontinuity</u>. This discontinuity occurs when a function is defined differently at a single point (as in Example 6.6), or when a function is not defined at a certain point (as in Example 6.2). These points are also referred to as removable discontinuities, as their limit usually exists, and if the point is removed from the domain, the function appears continuous (cf. Examples 6.2 and 6.6; Fig. 6.2).

2. <u>Jump discontinuity</u>. This type of discontinuity occurs when a function approaches two different values from the two sides of the discontinuity and can be thought of as a jump, as illustrated in Example 6.4. It often occurs for *piecewise* functions that are defined differently for different parts of their domain. Another example is the Heaviside step function (see Fig. 6.4) which can be defined as:

$$f(x) = \begin{cases} 0, & x \in (-\infty, 0) \\ \frac{1}{2}, & x = 0 \\ 1, & x \in (0, \infty) \end{cases}$$

3. <u>Asymptotic discontinuity</u>. To understand this type of discontinuity you have to understand what an asymptote is. This is a value for an argument where its function approaches $\pm\infty$ (vertical asymptote) or a value that a function approaches when its argument goes to $\pm\infty$ (horizontal asymptote). When a function has a vertical asymptote, it has an asymptotic discontinuity for that argument, as in Example 6.8, where the function has an asymptotic discontinuity at $x = 0$. Note that this function also has a horizontal asymptote at $y = 0$, as can be observed from Fig. 6.3. Similarly, as an example, the function $f(x) = \frac{x^2+3}{x-1}$ has an asymptotic discontinuity at $x = 1$.

6.7 Special Limits

We here provide some special limits without proof.

Box 6.2 Special limits

1. $\lim\limits_{x \to \infty} \left(1 + \frac{1}{x}\right)^x = e$ or $\lim\limits_{x \to 0} (1 + x)^{\frac{1}{x}} = e$

(continued)

Box 6.2 (continued)

2. $\lim\limits_{x \to 0} \frac{\sin x}{x} = 1$

3. $\lim\limits_{x \to 0} \frac{e^x - 1}{x} = 1$

4. $\lim\limits_{x \to 0} \frac{\ln(1+x)}{x} = 1$

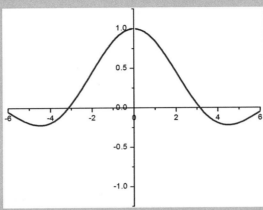

The graph of $\frac{\sin x}{x}$ is plotted here for illustration. The reason that its limit is equal to 1 for x approaching 0 is that for very small x ($|x| \ll 1$), $\sin x \approx x$.

6.8 Derivatives

Derivatives are very important for many branches of science in which the rate of change of continuous functions has to be determined. Whenever you need to determine changes over time or growth rates, or when calculating velocity or acceleration, or maximum or minimum values of a function, you use derivatives. In formally defining derivatives, we will use limits again.

Intuitively, we can think of a derivative as the amount by which a function changes when its argument changes an arbitrarily small amount. The value of the derivative depends on the value of the argument. A simple example is the velocity of a car: its velocity is determined by its displacement in a given period of time (change in position divided by change in time) and is equal to the derivative of its position as a function of time, with respect to time.

It turns out that the derivative is equal to the slope of the *tangent* of the function (cf. Fig. 6.1). The tangent (or tangent line) is the line that just touches the graph and is maximally parallel to the graph of the function. Below (Fig. 6.5) are two examples.

It can clearly be seen for the left graph in Fig. 6.5 that the slope of the tangent line changes for different values of x (x_1 and x_2). But the slope of the tangent line (derivative) does not always change as a function of x, as can be observed for the right graph in Fig. 6.5. For this linear function, the tangent line is the same as the function, and it is the same in both points (and actually everywhere). Indeed, for any linear function the derivative is a constant.

We have just introduced the derivative of functions by simple examples and hopefully you have gained an intuitive idea about derivatives. Now, let's consider the formal definition of a derivative, which employs limits that were just introduced to you.

Let's start by considering the left graph in Fig. 6.5 and the points on the graph for its arguments x_1 and x_2. When x_1 and x_2 are relatively far apart, the straight line between them looks as in Fig. 6.6 (*left*). When we draw the same line for x_1 and x_2 that are closer together, it looks like a tangent (Fig. 6.6 (*right*)).

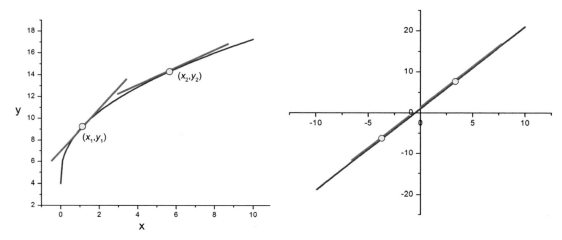

Fig. 6.5 Examples of tangent lines (*red*) for two different functions (their graphs are plotted in *blue*), for different values of their arguments (indicated with *circles*). Left: $y = \left(\sqrt[3]{x}+2\right)^2$, right: $y=2x+1$.

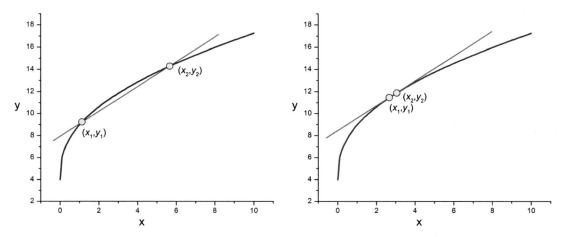

Fig. 6.6 Same graph as in Fig. 6.5 (*left*). Left: Plot of the straight line (*red*) through the two points (x_1,y_1) and (x_2,y_2) on the curve when these points are far apart. Right: The same situation when (x_1,y_1) and (x_2,y_2) are brought closer together.

If this line is given by the equation

$$g(x) = ax + b$$

then the slope a of this line through the two points (x_1, y_1) and (x_2, y_2) is given by:

$$a = \frac{y_2 - y_1}{x_2 - x_1}$$

This is also the rate of change of the function y with respect to its input variable x between x_1 and x_2. If the two points get closer and closer together (which should remind you of the definition of a limit), then the slope a is equal to the slope of the tangent at that point:

$$\lim_{x_2 \to x_1} \frac{y_2 - y_1}{x_2 - x_1}$$

This is the instantaneous rate of change of the function at $x = x_1$. We can also rewrite this to:

$$\lim_{\Delta x \to 0} \frac{\Delta y}{\Delta x}$$

where Δ (delta) is a change, with $\Delta y = y_2 - y_1$ and $\Delta x = x_2 - x_1$. We now arrive at the formal definition of a derivative:

$$\frac{dy}{dx} = \lim_{\Delta x \to 0} \frac{\Delta y}{\Delta x} \tag{6.1}$$

Alternatively, the derivative $\frac{df(x)}{dx}$ of a function $f(x)$ with respect to x is written as $f'(x)$. For the special case of derivation with respect to time, yet another (dot) notation is reserved: $\frac{df(t)}{dt} = \dot{f}$. Note that the act of calculating a derivative is referred to as differentiation.

Box 6.3 summarizes different notations for derivatives.

Box 6.3 Alternative definitions of a derivative

When y is a function of one variable:

$$\frac{dy}{dx} = \frac{df(x)}{dx} = \frac{d}{dx}f(x) = \frac{dy}{dx} = y'$$

6.9 Basic Derivatives and Rules for Differentiation

Before any examples, I will here provide the derivatives for several basic functions (Table 6.1). Together with the rules for differentiation in Table 6.2, they will allow you to differentiate most functions.

Thus, the derivative of a constant function is zero, the derivative of a function of x to some power is always proportional to that same function for which the power is lowered by one. We will see several examples of this below. For now, just try to determine the derivative of $f(x) = \frac{1}{x}$, applying this rule. What is also interesting is that the derivative of an exponential function turns out to be that same exponential. This is an important property of derivatives, that we will make use of in the next chapter on integrals. The exponential is useful to describe many physical phenomena, most importantly waves. Since sound, light and alternating current can all be described in terms of waves, the exponential is used to describe not only the *propagation*, but also the velocity of sound and light.

Table 6.1 Derivatives $\frac{dy(x)}{dx}$ for basic functions $y(x)$

$y(x)$	$\frac{dy(x)}{dx}$
C (constant)	0
$x^n, \quad n \in \mathbb{Q}$	nx^{n-1}
e^x	e^x
$\sin x$	$\cos x$
$\cos x$	$-\sin x$
$\log_a x$	$\frac{1}{x \ln a}, (x > 0, a > 0 \text{ and } a \neq 1)$
$\ln x$	$\frac{1}{x}, \quad (x > 0)$
a^x	$a^x \ln a$

Table 6.2 Combined functions and rules for their differentiation

$y(x)$	$\frac{dy(x)}{dx}$
$cu(x), c$ is a constant	$cu'(x)$
$u(x) + v(x)$	$u'(x) + v'(x)$
$u(x)v(x)$	$u'(x)v(x) + u(x)v'(x)$ product rule
$\frac{u(x)}{v(x)}$	$\frac{u'(x)v(x) - u(x)v'(x)}{v^2(x)}$ quotient rule
$y(x) = y(u) \quad u = u(x)$	$\frac{dy}{du}\frac{du}{dx}$ chain rule
$y = y(v) \quad v = v(u) \quad u = u(x)$	$\frac{dy}{dx} = \frac{dy}{dv}\frac{dv}{du}\frac{du}{dx}$ chain rule

As an example, combining the rule for differentiation of polynomials with the chain rule, we can derive that for $y(x) = \frac{1}{v(x)}$, $y'(x) = -\frac{v'(x)}{v^2(x)}$. Also, the chain rule can be applied multiple times (Table 6.2, last row), so that the derivative of $y(x)=y(v)$, where $v=v(u)$ and $u=u(x)$ is $y'(x) = \frac{dy}{dv}\frac{dv}{du}\frac{du}{dx}$.

Now let's practice with the rules in Tables 6.1 and 6.2 by means of some examples.

Example 6.11

Calculate the derivative of the function $y(x)=2x-3\cos x$.
  Since $y(x)$ is a sum of two functions $2x$ and $-3\cos x$, we can calculate its derivative by calculating the separate derivatives of the two functions, according to the rule for a sum of functions in Table 6.2:

$$y' = (2x)' + (-3\cos x)'$$

Then, we use the rule for multiplication with a constant in the same table:

$$y' = 2x' + (-3)(\cos x)'$$

And now we can calculate the derivative using the rules in Table 6.1:

$$y' = 2 \cdot 1 + (-3)(-\sin x) = 2 + 3\sin x$$

Example 6.12

Calculate the derivative of the function $y(x)=x^4+3x$.
  We can again apply the rules in Tables 6.1 and 6.2:

$$y' = (x^4)' + (3x)' = 4x^{4-1} + 3(x)' = 4x^3 + 3$$

Example 6.13

Calculate the derivative of the function $y(x)=\log_3 x$.
  Again, we will use the rules in Table 6.1:

$$y' = \frac{1}{x \ln 3}$$

Example 6.14

Calculate the derivative of the function $y(x)=\ln x+0.5\sin x$.
  Applying the rules in Tables 6.1 and 6.2 yields:

$$y' = (\ln x)' + 0.5(\sin x)' = \frac{1}{x} + 0.5\cos x$$

Example 6.15

Calculate the derivative of the function $y(x) = \sin 4x$.
   Here we have to apply the chain rule in Table 6.2, where $u(x) = 4x$ and $y(u) = \sin u$:

$$y'(x) = \frac{d(\sin u)}{du} \frac{d(4x)}{dx} = (\cos 4x) \cdot 4 = 4\cos 4x$$

Exercise

6.4. Differentiate the following functions:

   a) $y(x) = 3x^2 + 2x^3 - 1$
   b) $y(x) = \sin 3x$
   c) $y(x) = 17e^{3t} + x^{22}$ (note that t is a constant)
   d) $y(x) = 7x$
   e) $y(x) = 33\cos 11x + 5$
   f) $y(x) = e^{-x}$
   g) $y(x) = 3x^4 - \sqrt{x}$

To prepare you for the next set of exercises, we provide some more examples first.

Example 6.16

Calculate the derivative of the function $y(x) = x\sin x$.
   We apply the product rule in Table 6.2, where $u(x) = x$ and $v(x) = \sin x$:

$$y'(x) = (x\sin x)' = (x)'\sin x + x(\sin x)' = \sin x + x\cos x$$

Example 6.17

Calculate the derivative of the function $y(x) = x^k a^{nx}$, where a, k and n are constants.
   We again apply the product rule and the chain rule in Table 6.2:

$$y'(x) = \left(x^k\right)' a^{nx} + x^k (a^{nx})' = kx^{k-1}a^{nx} + x^k na^{nx}\ln a = x^{k-1}a^{nx}(k + nx\ln a)$$

Example 6.18

Calculate the derivative of the function $y(x) = \frac{\sin x + \cos x}{1 + \tan x}$.
   Now we use the quotient rule in Table 6.2:

$$y'(x) = \frac{(\sin x + \cos x)'(1 + \tan x) - (\sin x + \cos x)(1 + \tan x)'}{(1 + \tan x)^2}$$

(continued)

Example 6.18 (continued)

We now need the derivative of tan$x$, which we just provide here. However, it is a good exercise to calculate it yourself using the definition $\tan x = \frac{\sin x}{\cos x}$ and the quotient rule in Table 6.2.

$$(\tan x)' = \frac{1}{\cos^2 x}$$

We can now continue calculating the derivative of $y(x)$ using the basic rules in Table 6.1:

$$y'(x) = \frac{(\cos x - \sin x)(1 + \tan x) - (\sin x + \cos x)\frac{1}{\cos^2 x}}{(1 + \tan x)^2}$$

$$= \frac{\cos x - \sin x + \sin x - \frac{\sin^2 x}{\cos x} - \frac{\sin x}{\cos^2 x} - \frac{1}{\cos x}}{(1 + \tan x)^2}$$

$$= \frac{\cos x - \frac{1}{\cos x}(\sin^2 x + 1) - \frac{\sin x}{\cos^2 x}}{(1 + \tan x)^2}.$$

Exercise

6.5. Differentiate the following functions:

a. $2\sin\frac{x}{2}\cos\frac{x}{2}$. To simplify the final solution, use the formulas $\sin\frac{x}{2} = \sqrt{\frac{1}{2}(1 - \cos x)}$, $\cos\frac{x}{2} = \sqrt{\frac{1}{2}(1 + \cos x)}$, which you can derive from the formula $\cos 2x = \cos^2 x - \sin^2 x$ (see Chap. 3).

b. $y = e^x \sqrt[4]{x}$

c. $y = a^x x^a$

d. $y = (x^2 - 1)\sqrt{x}$

e. $y = 3\sqrt[5]{x^2} - x\sqrt{x}$

f. $y = x \cdot 10^x$

g. $y = e^{ax}(a\sin x - \cos x)$

6.10 Higher Order Derivatives

So far we only addressed so-called first order derivatives, where we only calculate the derivative of a function once. In practice, higher order derivatives are also very common. The second order derivative is simply the derivative of the derivative of a function. In the same way you can calculate a third-order derivative - taking the derivative of the derivative of the derivative of a function. We summarized this in Box 6.4.

Box 6.4 Higher order derivatives

Second order derivative

$$y''(x) = y^{(2)}(x) = \frac{d^2y}{dx^2} = \frac{d}{dx}\left(\frac{dy}{dx}\right)$$

Third order derivative

$$y'''(x) = y^{(3)}(x) = \frac{d^3y}{dx^3} = \frac{d}{dx}\left(\frac{d^2y}{dx^2}\right) = \frac{d}{dx}\left[\frac{d}{dx}\left(\frac{dy}{dx}\right)\right]$$

For example, you already know that velocity is the (first order) derivative of displacement. Similarly, acceleration is the first order derivative of velocity, making it the second order derivative of displacement. Again, as acceleration is a second order derivative with respect to time, we can use the dot notation introduced earlier:

When $x(t)$ is the displacement function, velocity is given by $\dot{x} = \frac{dx}{dt} = v(t)$ and acceleration by $\ddot{x} = \frac{d^2x}{dt^2} = a(t)$.

Example 6.19

Calculate the first, second and third order derivatives for the function $y(x) = 3x^3 + 2x^2 - 1$.

This is done step by step, by first calculating the first order derivative, then the second and finally the third order derivative:

$$y'(x) = 9x^2 + 4x$$
$$y''(x) = 18x + 4$$
$$y'''(x) = 18$$

As you can see, the third order derivative of this function is constant.

Exercise

6.6. Calculate the first, second and third order derivatives for the following functions:

a. $y(t) = e^{-i\omega t}$, remember that $i \cdot i = -1$ (Sect. 1.2.4)
b. $y(x) = \sin 3x$
c. $y(x) = 17e^{3t} + x^{22}$
d. $y(x) = \frac{lnx}{x}$

6.11 Partial Derivatives

Partial derivatives of a function of two or more variables (e.g. $f(x,y,z)$) are derivatives with respect to one variable while the other variables are considered constant. Partial derivatives are often used to determine (spatial) gradient changes in different directions (e.g in fMRI or diffusion tensor imaging (DTI)). We will first explain partial derivatives using an example.

Example 6.20

Consider the function $f(x,y)=3x^2+5xy+y^3$. Calculate the partial derivative with respect to x, denoted by $\frac{\partial f(x,y)}{\partial x}$ (see Box 6.4 for notation of partial derivatives).

In this case, variable y is considered constant. The rules for partial differentiation are the same as the rules for general differentiation. The partial derivative with respect to x is thus:

$$\frac{\partial f(x,y)}{\partial x} = \frac{\partial}{\partial x}\left(3x^2\right) + \frac{\partial}{\partial x}(5xy) + \frac{\partial}{\partial x}\left(y^3\right) = 6x + 5y + 0 = 6x + 5y$$

The last term y^3 was constant in x, and the derivative of a constant is 0.

Box 6.5 Partial derivative notation

For a function of two variables $f(x,y)$:

$$\frac{\partial f(x,y)}{\partial x} = \left(\frac{df(x,y)}{dx}\right)_{y=constant}$$

$$\frac{\partial f(x,y)}{\partial y} = \left(\frac{df(x,y)}{dy}\right)_{x=constant}$$

More notations for partial derivatives:

$$\frac{\partial f(x,y,..)}{\partial x} = \frac{\partial}{\partial x} f(x,y,..) = f_x(x,y,\ldots) = \partial_x f(x,y,\ldots)$$

Similar to higher order derivatives of a function with one variable, there are second and higher order partial derivatives. Note that, in case of partial derivatives, you can first calculate the partial derivative with respect to one variable (say x) and then with respect to the second variable (say y) (Box 6.6).

Box 6.6 Second order partial derivatives

$$\frac{\partial^2 f(x,y,\ldots)}{\partial x^2} = \frac{\partial}{\partial x}\left(\frac{\partial f(x,y,..)}{\partial x}\right)$$

$$\frac{\partial^2 f(x,y,\ldots)}{\partial y^2} = \frac{\partial}{\partial y}\left(\frac{\partial f(x,y,..)}{\partial y}\right)$$

$$\frac{\partial^2 f(x,y,\ldots)}{\partial x \partial y} = \frac{\partial}{\partial x}\left(\frac{\partial f(x,y,..)}{\partial y}\right)$$

Note that $\frac{\partial^2 f(x,y,\ldots)}{\partial x \partial y} = \frac{\partial^2 f(x,y,\ldots)}{\partial y \partial x}$ for continuous functions.

Example 6.21

Consider the function from the previous example

$$f(x, y) = 3x^2 + 5xy + y^3$$

and calculate all possible second order derivatives as in Box 6.6.

$$\frac{\partial^2 f(x, y)}{\partial x^2} = \frac{\partial^2}{\partial x^2}(3x^2) + \frac{\partial^2}{\partial x^2}(5xy) + \frac{\partial^2}{\partial x^2}(y^3) = 6 + 0 + 0 = 6$$

$$\frac{\partial^2 f(x, y)}{\partial y^2} = \frac{\partial^2}{\partial y^2}(3x^2) + \frac{\partial^2}{\partial y^2}(5xy) + \frac{\partial^2}{\partial y^2}(y^3) = 0 + 0 + 6y = 6y$$

$$\frac{\partial^2 f(x, y)}{\partial x \partial y} = \frac{\partial^2}{\partial x \partial y}(3x^2) + \frac{\partial^2}{\partial x \partial y}(5xy) + \frac{\partial^2}{\partial x \partial y}(y^3) = 0 + 5 + 0 = 5$$

Thus, calculating partial derivatives is easy as long as you know how to calculate derivatives in general.

Exercise

6.7. Calculate all possible first and second order partial derivatives of the function: $f(x, t) = e^{-ixt}$

6.12 Differential and Total Derivatives

In Sect. 6.8 we defined the derivative as

$$\frac{dy}{dx} = \lim_{\Delta x \to 0} \frac{\Delta y}{\Delta x}$$

Here $\Delta x = (x_2 - x_1)$ is a difference between two points. Thus, for very small Δx we can think of it as being equal to dx; this is then referred to as a differential. Thus a differential is an infinitesimally small (very, very small) difference in the variable. We will need the differential for understanding the total derivative (below) and integrals (Chap. 7). Once we accept this definition of a differential we can manipulate differential formulas as if the differentials are variables on their own. For example, as

$$\frac{df(x)}{dx} = \frac{df(x)}{dx}$$

We can rewrite this to:

$$df(x) = \frac{df(x)}{dx} dx$$

More generally, for a function *f(x,y)* the total derivative is equal to:

$$df(x,y) = \frac{\partial f(x,y)}{\partial x} dx + \frac{\partial f(x,y)}{\partial y} dy$$

Example 6.22

Calculate the total derivative of $f(x,t) = e^{-ixt}$.
   We already know from Exercise 6.7 that $\frac{\partial f(x,t)}{\partial x} = -ite^{-ixt}$ and $\frac{\partial f(x,t)}{\partial t} = -ixe^{-ixt}$.
Thus:

$$df(x,t) = \frac{\partial f(x,t)}{\partial x} dx + \frac{\partial f(x,t)}{\partial t} dt = -ite^{-ixt} dx + -ixe^{-ixt} dt$$

6.13 Practical Use of Derivatives

We already mentioned that derivatives are widely used in science. For example, fitting a function to a given dataset requires derivatives, because derivatives are used to calculate maxima and minima. Let's have a closer look at how derivatives are used to calculate extrema of a function.

6.13.1 Determining Extrema of a Function

For a smooth function (a function that has well-defined derivatives across its entire domain), its extrema (minima and maxima) can be determined by identifying the points where the function is flat (where its slope is zero), or, in other words, where its derivative equals zero.
   Consider the function in Fig. 6.7. It is a quadratic function with one minimum. But how do we know where the minimum is exactly? We find it by taking the first derivative with respect to *x* and determining its root. Thus, for this function

$$f(x) = (x+3)^2 + 1$$

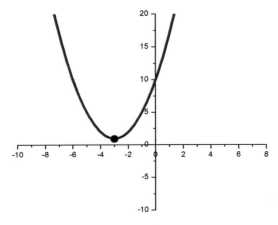

Fig. 6.7   Quadratic function $(x+3)^2+1$ with a minimum at $(x,y)=(-3,1)$.

we solve the equation

$$\frac{df(x)}{dx} = 0$$

To determine the derivative, we apply the chain rule in Table 6.2, using that $u(x)=x+3$, resulting in:

$$\frac{df(x)}{dx} = 2(x+3)$$

Next, we find the root of the derivative:

$$2(x+3) = 0$$

$$(x+3) = 0$$

Thus, the function has a minimum at $x=-3$.

How can we determine whether it is a minimum or maximum? The easiest is to look at the points around it and see whether the function is higher or lower than this extreme value, but we can also find out mathematically (Box 6.7).

Box 6.7 Distinguishing maxima and minima of a function

1) Determine the *stationary points* x_i of the function $f(x)$ by finding the roots of its derivative.
2) The stationary points are *local* extrema (maxima or minima) when $f'(x_i)=0$ and $f''(x_i)\neq0$.
3) The local extrema is a local maximum when $f''(x_i)<0$ and a local minimum when $f''(x_i)>0$.
4) When $f''(x_i)=0$ the nature of the stationary point must be determined by other means, often by noting a sign change around that point.

Example 6.23

Consider the function $f(x)=x^2+5x+4$ and determine its stationary points.

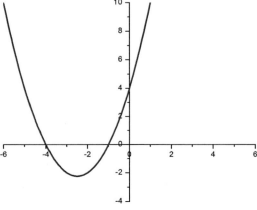

(continued)

Example 6.23 (continued)

To determine whether this function has extrema, we first calculate its first derivative and find its roots:

$$f'(x) = 2x + 5 = 0$$

Thus, for $x_1 = -\frac{5}{2}$ this function has a stationary point. But is it an extrema, and if so is it a maximum or a minimum? To determine this, we consider the second derivative of f and determine its sign at x_1:

$$f''(x) = 2$$

The second order derivative is constant and positive for the entire domain. Thus, $f''(x_1) > 0$. If we compare this result to the rules in Box 6.6, we find that the function has a minimum at $x_1 = -\frac{5}{2}$. We can confirm this by studying the graph of this function in the figure.

Example 6.24

Consider the function $f(x) = x^3$ and determine its stationary point(s). Are they extrema?
First we determine the roots of the derivative of f:

$$f'(x) = 3x^2 = 0$$

This equation has a solution for $x_1 = 0$, i.e. $x_1 = 0$ is a stationary point of the function. Next, we calculate the second order derivative and determine its sign at $x_1 = 0$:

$$f''(x) = 6x$$

Hence, at $x_1 = 0$ the second order derivative $f''(x_1) = 4x_1 = 0$. Thus, f has a stationary point but not an extrema. Let's inspect the function's graph in the figure to find out why. The function has neither a maximum nor a minimum at $x_1 = 0$, but it bends. This is a point of inflection. Not all third order polynomial functions apparently have a maximum or a minimum.

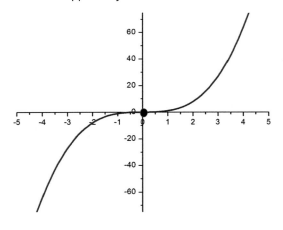

Example 6.25

Consider the function $f(x)=2\cos x - x$ and determine its stationary point(s). Are they extrema?

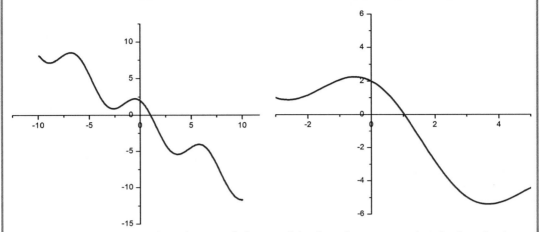

If we study the graph of this function (left part of the figure) we can see that the function has several local maxima and minima but none is an *absolute* maximum or minimum; an extrema over the entire domain of the function. If we consider only the limited domain $x \in (-3, 5)$ then the function has exactly one minimum and one maximum (right part of the figure). We find the locations of these extrema by finding the roots of the derivative of the function (you need a calculator for this):

$$f'(x) = 0 \text{ when } \sin x = -\frac{1}{2} \text{ and } x \in (-3, 5) \;\; \rightarrow x_{max} = -0.52 \text{ and } x_{min} = 3.66$$

Thus not every function has an absolute maximum or minimum. Some functions have many local maxima or minima, some have points of inflection. This is very important to realize as the maximum or minimum are used when finding the best fit of measured data to a certain function (or distribution).

6.13.2 (Linear) Least Squares Fitting

A frequently used practical application of finding a minimum is (linear) least squares fitting. Least squares fitting is a method to fit a function $f(x, y, \ldots)$ to a given dataset. The idea is to minimize the offsets between the data and the fitting function. The fitting function is usually chosen based on some a priori hypothesis about relationships in the data, or experience, and it can be of any form such as linear, exponential or a power function. Here, to demonstrate the use of derivatives, we will explain least squares fitting of a linear function.

The general form of a linear function (straight line) is:

$$f(x) = ax + b \tag{6.2}$$

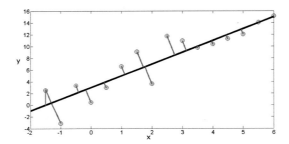

Fig. 6.8 Data points (*blue*) fitted by a linear function (*black*). Residuals calculated as perpendicular distances (*green*). For one point we illustrate how residuals can be calculated as vertical distances (*red line*).

where a and b are constants. If we have measured a variable y, at specific points x, we have a collection of data pairs (x_i, y_i) where $i = 1, \ldots, n$. Here n is the number of data pairs. The y_i could be, e.g., temperature read-outs at time points x_i. Our aim is to fit the linear function in Eq. 6.2 to the data. The least squares fitting procedure will aim to minimize the sum of the square residuals. Here, the residual R_i at a point x_i is the difference between the measured temperature y_i and the fitted function at x_i:

$$R_i = y_i - f(x_i)$$

In practice, the perpendicular distances are more often used as residuals (see Fig. 6.8). For the linear fitting function in Eq. 6.1 they are given by:

$$d_i = \frac{|y_i - (ax_i + b)|}{\sqrt{1 + b^2}}$$

The least squares fitting procedure aims to minimize the sum of squares of these distances. The sum of squares is used because the derivative of the square function exists in all points, allowing differentiation of the residuals in all points. The square residuals are given by:

$$d_i^2 = \left(\frac{|y_i - (ax_i + b)|}{\sqrt{1 + b^2}} \right)^2 = \frac{(y_i - (ax_i + b))^2}{1 + b^2}$$

The sum of squares is then given by:

$$R^2 = \sum_{i=1}^{n} \frac{(y_i - (ax_i + b))^2}{1 + b^2}$$

To minimize it we of course use derivatives! We find the minimum of this sum of squares of residuals by solving the system of equations:

$$\frac{\partial y}{\partial a}\left(R^2\right) = 0 \ \text{ and } \ \frac{\partial y}{\partial b}\left(R^2\right) = 0$$

6.13.3 Modeling the Hemodynamic Response in Functional MRI

In neuroimaging, we try to determine several brain characteristics, such as its anatomy, the distribution of white matter tracts and the gray matter volume. We also use neuroimaging techniques to determine what happens in the brain when a person performs some task (like reading, or solving a puzzle). Functional neuroimaging is used to accomplish this; when using MRI techniques we call this functional MRI (fMRI). fMRI is a great technique to measure brain activation when a person is performing a task, because it has such high spatial resolution (up to $2 \times 2 \times 2 \ \text{mm}^3$). However, the actual brain activation is not measured directly with fMRI; instead the change in *blood oxygenation level*, in response to a change in neuronal electrical activity due to task performance, is measured. This human *hemodynamic response* to brief periods of neural activity is sluggish and therefore delayed and dispersed in time. Approximately 5–6 s after the onset of neuronal activity, the blood oxygenation level rises to its maximum (see Fig. 6.9 top panel). We already mentioned in Sect. 4.3.2 Example 4.4, that the task-related brain activation can be modeled by a *regressor*, that contains a prediction of the brain response. For an experiment that is designed in blocks of alternating rest and task performance (block design), we expect the brain response to alternate in areas that are involved in the task, as well. To predict a response, i.e. to make a regressor, for fMRI experiments, we convolve the regressor that contains task information

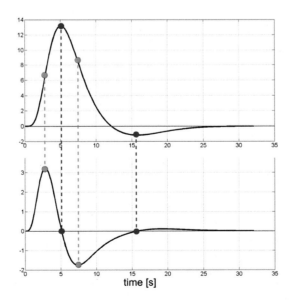

Fig. 6.9 *Top*: model of the hemodynamic response function, *bottom*: derivative of the function in the *top panel*. *Red dots* indicate extrema of the *top* function, that correspond to roots of the *bottom* function. *Blue dots* indicate points of inflection of the *top* function, that correspond to extrema of the *bottom* function.

(onsets and the duration of each block, where rest is indicated by 0 and task by 1 for a block design) and a prediction of the hemodynamic response. Convolution is explained in Sect. 7.5.2. In fMRI this prediction of the hemodynamic response is called the hemodynamic response function (HRF) (see Fig. 6.9 top panel).

To allow for more flexibility in the shape of the prediction of this response, the first temporal derivative of the HRF (Fig. 6.9 bottom panel) is also used to convolve the block design regressor with (see explanation in Liao et al. 2002 or Henson et al. 2002), giving a second regressor. Note how the derivative of the HRF appears shorter-lived. Also, this is a nice example showing that where a function has an extrema (red dots in Fig. 6.9 top panel) its derivative is zero. One can also spot the points of inflection (blue dots in Fig. 6.9 top panel), that are of a different nature than the ones in Example 6.24. At its points of inflection, the original function (the HRF) does not have stationary points, but its derivative has an extrema.

6.13.4 Dynamic Causal Modeling

Neuroscientists are interested in how the brain functions and particularly, in how brain regions communicate with each other. One method to investigate brain communication is dynamic causal modeling or DCM (Friston et al., 2003). DCM employs derivatives to model and investigate the dynamic properties of local brain activity. In dynamic causal modeling bilinear equations are used to describe the interaction between two brain regions.

To explain this in more detail, let's consider two brain regions 1 and 2, with their activities $z_1(t)$ and $z_2(t)$, as measured during a neuroimaging experiment in which participants had to watch images with positive or negative content. Presentation of images with a positive content is modeled as $u_p(t)$ (which is 1 when a positive image is shown and 0 otherwise) and similarly $u_n(t)$ is used to model the presentation of negative images. In DCM one assumes a causal model of interactions between brain regions and how these interactions change due to experimental manipulations. For example, you can assume that activity in region 2 causes activity in region 1 and that that activity will be affected by watching negative images (as modeled by $u_n(t)$; Fig. 6.10a). Or you can assume that the activities in both regions affect each other and that the activity in region 1 is affected by watching positive images (as modeled by $u_p(t)$; Fig. 6.10b). These assumptions result in two different causal models.

In general, the dynamic causality is modeled as:

$$\frac{dz(t)}{dt} = Az(t) + B^j u_j(t)z(t) + Cu_j$$

In the example above, $z(t)$ is the activity in either region 1 or region 2. The u_j are the inputs $u_p(t)$ and $u_n(t)$. The coefficients A, B^j and C need to be determined. What one can observe is that the change in activity in one brain region $(\frac{dz(t)}{dt})$ is coupled via coefficient A to activity in another brain region. We say that activity in the latter region causes the change in activity in the first region, where the amount of coupling is given by A. The extent to which the

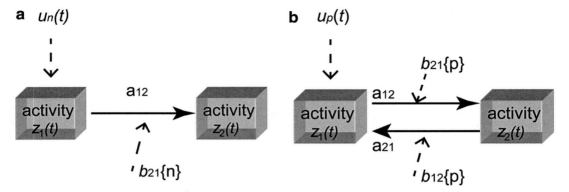

Fig. 6.10 DCM models to model brain activity $z_1(t)$ and $z_2(t)$ in two brain regions during a neuroimaging experiment in which participants watched images with positive or negative content. Presentation of images with a positive content is modeled by $u_p(t)$ and $u_n(t)$ is used to model the presentation of negative images. Different assumptions result in different models. (a) DCM model of activity in region 2 causing activity in region 1, assuming that activity in region 1 will be affected by watching negative images; (b) DCM model of activities in both regions affecting each other, assuming that the activity in region 1 is affected by watching positive images.

coupling is affected by inputs (u_j) is modeled by coefficients B. The direct effect of these inputs on brain activity is modeled by coefficients C.

Example 6.26

Determine the equations that describe the causal model in Fig. 6.10a.

Since we have only one causal interaction from region 1 to region 2 we need only one dynamic causal equation to describe the model:

$$\frac{dz_2(t)}{dt} = a_{12}z_1(t) + b_{12}^n u_n z_1(t)$$

However, the input u_n also directly affects activity in brain region 1, so we need another equation to describe that:

$$\frac{dz_1(t)}{dt} = c_1^n u_n(t)$$

Example 6.27

Determine the equations that describe the causal model in Fig. 6.10b.

These equations can be found in a similar way as for Example 6.26. This model is a bit more complex, but can nevertheless be described by two equations:

$$\frac{dz_2(t)}{dt} = a_{12}z_1(t) + b_{12}^p u_p z_1(t)$$

$$\frac{dz_1(t)}{dt} = a_{21}z_2(t) + b_{21}^p u_p z_2(t)c_1^p u_p(t)$$

Glossary

Absolute As in absolute maximum or minimum: the largest maximum or minimum value over the entire domain of a function.

Argument Input variable of a function.

Asymptote A line or a curve that approaches a given curve arbitrarily closely; their point of touch tends towards infinity. It is the tangent of a curve at infinity.

Asymptotic discontinuity A function that has a vertical asymptote for the argument belonging to the discontinuity.

Blood oxygenation level The level of oxygen in the blood.

Continuous A function that is defined on its domain, for which sufficiently small changes in the input result in arbitrarily small changes in the output.

Derivative Rate of change of a function; also the slope of the tangent line.

Differential Infinitesimal differences of a function.

Discrete Opposite of continuous; a discrete variable can only take on specific values.

Domain The set of arguments for which a function is defined.

Extrema Collective name for maxima and minima of a function.

Hemodynamic response Here: the increase in blood flow to active brain neuronal tissue.

Infinity An abstract concept that in the context of mathematics can be thought of as a number larger than any number.

Limit A limit is the value that a function or sequence "approaches" as the variable approaches some value.

Local extrema As in local maximum or minimum: a maximum or minimum value of the function in a neighbourhood of the point.

One-sided limit Limit that only exists for the variable of a function approaching some value from one side.

Optimisation problem The problem of finding the best solution from all possible solutions.

Partial derivative A derivative of a function of several variables with respect to one variable, considering remaining variables as a constant.

Piecewise A function that is defined differently for different parts of its domain.

Point of inflection Point where the curvature changes sign, i.e. where the derivative has an extrema.

Propagation Movement.

Regressor An independent variable that can explain a dependent variable in a regression model.

Removable discontinuity (Also known as point discontinuity) discontinuity that occurs when a function is defined differently at a single point, or when a function is not defined at a certain point.

Root Point where a function is equal to zero.

(To) sample To digitize an analog signal, analog-to-digital (A/D) conversion.

Stationary point A point on a curve where the derivative is equal to zero.

Tangent (line) A straight line that touches a function at only one point, also an instantaneous rate of change at that point.

Symbols Used in This Chapter (in Order of Their Appearance)

$\lim_{x \to a} f(x)$	Limit
$\lim_{x \to a^+} f(x)$ or $\lim_{x \downarrow a} f(x)$	Limit from the right
$\lim_{x \to a^-} f(x)$ or $\lim_{x \uparrow a} f(x)$	Limit from the left

$\dfrac{dy}{dx} = y'(x)$	(First order) derivative
$\dfrac{df(t)}{dt} = \dot{f}$	Time derivative
$y''(x) = y^{(2)}(x)$	Second order derivative
$y'''(x) = y^{(3)}(x)$	Third order derivative
$\dfrac{\partial\, f(x,y)}{\partial\, x}$	Partial derivative with respect to x
$\dfrac{\partial\, f(x,y)}{\partial\, y}$	Partial derivative with respect to y
$df(x,y)$	Differential

Overview of Equations for Easy Reference

Limit

$$\lim_{x \to a} f(x) = L$$

Function with multiple domains

$$f(x) = \begin{cases} f_1(x), & \text{for } x \in (-\infty, a] \\ f_2(x), & \text{for } x \in (a, b] \\ f_3(x), & \text{for } x \in (b, \infty) \end{cases}$$

Arithmetic rules for limits

1. $\lim\limits_{x \to c} a \cdot f(x) = a \cdot \lim\limits_{x \to c} f(x)$, when a is constant
2. $\lim\limits_{x \to c} [f(x) \pm g(x)] = \lim\limits_{x \to c} f(x) \pm \lim\limits_{x \to c} g(x)$
3. $\lim\limits_{x \to c} [f(x)g(x)] = \lim\limits_{x \to c} f(x) \cdot \lim\limits_{x \to c} g(x)$
4. $\lim\limits_{x \to c} \left[\dfrac{f(x)}{g(x)}\right] = \dfrac{\lim\limits_{x \to c} f(x)}{\lim\limits_{x \to c} g(x)}$, if and only if $\lim\limits_{x \to c} g(x) \neq 0$
5. $\lim\limits_{x \to c} f(x)^n = \left[\lim\limits_{x \to c} f(x)\right]^n, n \in \mathbb{R}$
6. $\lim\limits_{x \to c} a = a$, when a is constant
7. $\lim\limits_{x \to c} x = c$

Heaviside function

$$f(x) = \begin{cases} 0, & x \in (-\infty, 0) \\ \dfrac{1}{2}, & x = 0 \\ 1, & x \in (0, \infty) \end{cases}$$

Special limits

1. $\lim\limits_{x\to\infty} \left(1 + \frac{1}{x}\right)^x = e$ or $\lim\limits_{x\to 0} (1 + x)^{\frac{1}{x}} = e$

2. $\lim\limits_{x\to 0} \frac{\sin x}{x} = 1$

3. $\lim\limits_{x\to 0} \frac{e^x - 1}{x} = 1$

4. $\lim\limits_{x\to 0} \frac{\ln(1+x)}{x} = 1$

Definition of derivative

$$\frac{dy}{dx} = \lim\limits_{\Delta x \to 0} \frac{\Delta y}{\Delta x}$$

Alternative expressions for the derivative if $y = f(x)$

$$y' = \frac{dy}{dx} = \frac{df(x)}{dx} = \frac{d}{dx} f(x) = f'(x)$$

Basic derivatives

$$y(x) = C, \qquad\qquad \frac{dy(x)}{dx} = 0, C \text{ is constant}$$

$$y(x) = x^n, \quad n \in \mathbb{Q}, \qquad \frac{dy(x)}{dx} = nx^{n-1}$$

$$y(x) = e^x, \qquad\qquad \frac{dy(x)}{dx} = e^x$$

$$y(x) = \sin x, \qquad\qquad \frac{dy(x)}{dx} = \cos x$$

$$y(x) = \cos x, \qquad\qquad \frac{dy(x)}{dx} = -\sin x$$

$$y(x) = \log_a x, \qquad\qquad \frac{dy(x)}{dx} = \frac{1}{x \ln a}, (x > 0, a > 0 \text{ and } a \neq 1)$$

$$y(x) = \ln x, \qquad\qquad \frac{dy(x)}{dx} = \frac{1}{x}, \quad (x > 0)$$

$$y(x) = a^x, \qquad\qquad \frac{dy(x)}{dx} = a^x \ln a$$

Basic rules for differentiation

$$y(x) = cu(x), c \text{ is a constant} \qquad y'(x) = cu'(x)$$
$$y(x) = u(x) + v(x), \qquad y'(x) = u'(x) + v'(x)$$
$$y(x) = u(x)v(x), \qquad y'(x) = u'(x)v(x) + u(x)v'(x), \text{ product rule}$$
$$y(x) = \frac{u(x)}{v(x)}, \qquad y'(x) = \frac{u'(x)v(x) - u(x)v'(x)}{v^2(x)}, \text{ quotient rule}$$
$$y(x) = y(u)u = u(x), \qquad \frac{dydu}{dudx}, \text{ chain rule}$$
$$y = y(v) \ \ v = v(u) \ \ u = u(x), \qquad \frac{dy}{dx} = \frac{dy}{dv}\frac{dv}{du}\frac{du}{dx}, \text{ chain rule}$$

Higher order derivatives.
Second order derivative

$$y''(x) = y^{(2)}(x) = \frac{d^2y}{dx^2} = \frac{d}{dx}\left(\frac{dy}{dx}\right)$$

Third order derivative

$$y'''(x) = y^{(3)}(x) = \frac{d^3y}{dx^3} = \frac{d}{dx}\left(\frac{d^2y}{dx^2}\right) = \frac{d}{dx}\left[\frac{d}{dx}\left(\frac{dy}{dx}\right)\right]$$

Partial derivatives.
First order partial derivatives for a function of two variables:

$$\frac{\partial f(x,y)}{\partial x} = \left(\frac{df(x,y)}{dx}\right)_{y=constant}$$
$$\frac{\partial f(x,y)}{\partial y} = \left(\frac{df(x,y)}{dy}\right)_{x=constant}$$

Second order partial derivatives

$$\frac{\partial^2 f(x,y,\ldots)}{\partial x^2} = \frac{\partial}{\partial x}\left(\frac{\partial f(x,y,..)}{\partial x}\right)$$

$$\frac{\partial^2 f(x,y,\ldots)}{\partial y^2} = \frac{\partial}{\partial y}\left(\frac{\partial f(x,y,..)}{\partial y}\right)$$

$$\frac{\partial^2 f(x,y,\ldots)}{\partial x \partial y} = \frac{\partial}{\partial x}\left(\frac{\partial f(x,y,..)}{\partial y}\right)$$

More notations for partial derivatives

$$\frac{\partial f(x, y, ..)}{\partial x} = \frac{\partial}{\partial x} f(x, y, ..) = f_x(x, y, \ldots) = \partial_x f(x, y, \ldots)$$

Total differential of a function of two variables

$$df(x, y) = \frac{\partial f(x, y)}{\partial x} dx + \frac{\partial f(x, y)}{\partial y} dy$$

Answers to Exercises

6.1. a) 18
   b) 12
   c) 1.5
   d) Does not exist.

6.2. a) -1
   b) -5
   c) 5/6
   d) $-1/2$

6.3. a) ∞
   b) 0
   c) 0
   d) ½
   e) ∞

6.4. a) $y' = 6x + 6x^2$
   b) $y' = 3\cos 3x$
   c) $y' = 22x^{21}$
   d) $y' = 7$
   e) $y' = -363\sin 11x$
   f) $y' = -e^{-x}$
   g) $y' = 12x^3 - \frac{1}{2\sqrt{x}}$

6.5. a) $y' = \cos x$
   b) $y' = e^x\left(x^{\frac{1}{4}} + \frac{1}{4}x^{-\frac{3}{4}}\right)$
   c) $y' = a^x x^{a-1}(x\ln a + a)$
   d) $y' = \frac{5x^2 - 1}{2\sqrt{x}}$
   e) $y' = \frac{6}{5}x^{\frac{3}{5}} - \frac{3}{2}\sqrt{x}$

f) $y' = 10^x(1 + x \ln 10)$

g) $y' = e^{ax} \sin x (1 + a^2)$

6.6. a) $y'(t) = -i\omega e^{-i\omega t}, y''(t) = -\omega^2 e^{-i\omega t}, y'''(t) = i\omega^3 e^{-i\omega t}$

b) $y'(x) = 3 \cos 3x, y''(x) = -9 \sin 3x, y'''(x) = -27 \cos 3x$

c) $y'(x) = 22x^{21}, y''(x) = 462x^{20}, y'''(x) = 9240x^{19}$

d) $y'(x) = x^{-2}(-\ln x + 1), y''(x) = x^{-3}(2 \ln x - 3), y'''(x) = x^{-4}(-6 \ln x + 11)$

6.7. $\frac{\partial f(x,t)}{\partial x} = -ite^{-ixt}, \quad \frac{\partial f(x,t)}{\partial t} = -ixe^{-ixt}, \quad \frac{\partial^2 f(x,t)}{\partial x^2} = -t^2 e^{-ixt}, \quad \frac{\partial^2 f(x,t)}{\partial t^2} = -x^2 e^{-ixt},$

$\frac{\partial^2 f(x,t)}{\partial x \partial t} = -e^{-ixt}(i + xt).$

References

Online Sources of Information

https://www.mathsisfun.com/calculus/continuity.html

http://tutorial.math.lamar.edu/Classes/CalcI/OneSidedLimits.aspx

https://www.khanacademy.org/

https://www.math.brown.edu/utra/discontinuities.html

http://en.wikipedia.org/wiki/Tangent

http://tutorial.math.lamar.edu/Classes/CalcI/Tangents_Rates.aspx

http://mathworld.wolfram.com/LeastSquaresFittingPerpendicularOffsets.html

Papers

C.H. Liao, K.J. Worsley, J. Poline, J.A.D. Aston, G.H. Duncan, A.C. Evans, Estimating the Delay of the fMRI Response. NeuroImage **16**, 593–606 (2002)

R.N.A. Henson, C.J. Price, M.D. Rugg, R. Turner, K.J. Friston, Detecting Latency Differences in Event-Related BOLD Responses: Application to Words versus Nonwords and Initial versus Repeated Face Presentations. NeuroImage **15**, 83–97 (2002)

K.J. Friston, L. Harrison, W. Penny, Dynamic causal modelling. NeuroImage **19**, 1273–1302 (2003)

7

Integrals

Branislava Ćurčić-Blake

After reading this chapter you know:

- what an integral is,
- what definite and indefinite integrals are,
- what an anti-derivative is and how it is related to the indefinite integral,
- what the area under a curve is and how it is related to the definite integral,
- how to solve some integrals and
- how integrals can be applied, with specific examples in convolution and the calculation of expected value.

7.1 Introduction to Integrals

There are many applications of integrals in everyday scientific work, including data and statistical analysis, but also in fields such as physics (see Sect. 7.7). To enable understanding of these applications we will explain integrals from two different points of view. Several examples will be provided along the way to clarify both.

Firstly, integrals can be considered as the opposite from derivatives, or as 'anti-derivatives'. This point of view will lead to the definition of *indefinite* integrals. Viewing integrals as the opposite of derivatives reflects that by first performing integration and then differentiation or vice versa, you basically get back to where you started. In other words, integration can be considered as the *inverse* operation of differentiation. However, while it is possible to calculate or find the derivative for any function, determining integrals is not always as easy. In fact, many useful indefinite integrals are not solvable, that is, they cannot be given as an

B. Ćurčić-Blake (✉)
Neuroimaging Center, University Medical Center Groningen, Groningen, The Netherlands
e-mail: b.curcic@umcg.nl

© Springer International Publishing AG 2017
N. Maurits, B. Ćurčić-Blake, *Math for Scientists*, DOI 10.1007/978-3-319-57354-0_7

analytic expression! In those cases, *numerical integration* may sometimes help, but that topic is outside the scope of this book.

Secondly, integrals (of functions of one variable) can be considered as the area under a curve. This point of view will lead to the definition of *definite* integrals. Integration can also be performed for functions of multiple variables and we will only briefly touch upon this topic in this chapter.

7.2 Indefinite Integrals: Integrals as the Opposite of Derivatives

As we mentioned before, one way to think about integrals is as the opposite or the reverse of derivatives; some people like to think about integrals as anti-derivatives. In other words, by integration you aim to find out what $f(x)$ is, given its derivative $f'(x)$, or more formally

$$f(x) = \int f'(x)dx$$

Here, the symbol for the *indefinite* integral $\int$ is introduced. In contrast to the definite integral that will be introduced in Sect. 7.3 the integral is here defined for the entire domain of the function. An important part of the integral is dx, the differential of the variable x. It denotes an infinitesimally small change in the variable (see Sect. 6.12), and shows that the variable of integration is x. The meaning of dx will become more clear when we explain definite integrals in Sect. 7.3.

Example 7.1

If $f'(x) = nx^{n-1}$, what is $f(x)$?

We thus need to find $f(x) = \int f'(x)dx = \int nx^{n-1}dx$. Since we know that $\frac{d}{dx}x^n = nx^{n-1}$, for this example we find that $f(x) = x^n$.

For ease of notation, we denote $F(x)$ as the integral of a function $f(x)$:

$$F(x) = \int f(x)dx$$

The function $f(x)$ that is integrated is also referred to as the *integrand*.

7.2.1 Indefinite Integrals Are Defined Up to a Constant

Since the derivative of a constant is zero, indefinite integrals are only defined up to a constant. This means that in practice, after finding the anti-derivative (also known as the *primitive*) of a function, you can add any constant to this anti-derivative and it will still fulfill the requirement that its derivative is equal to the function you were trying to find the anti-derivative for. An intuitive understanding of this property of indefinite integrals is provided by Example 7.2 and Fig. 7.1.

Example 7.2

If $f(x)=x^2$, then $f'(x)=2x$, but $f'(x)=2x$ is also true for $f(x)=x^2+3$ or $f(x)=x^2-5$. Figure 7.1 helps to understand this more intuitively: by adding a constant to a function of x, the function is shifted along the y-axis, but otherwise does not change shape. Hence, the derivative (the slope of the black lines in Fig. 7.1, see also Sect. 6.8) for a certain value of x, remains the same.

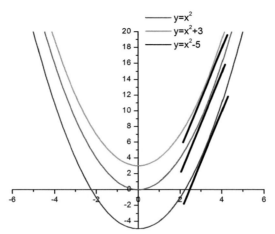

Fig. 7.1 The function $f(x)=x^2$ is plotted when different constants are added. It illustrates that the tangent at a specific value of x (*black lines*) has the same slope for all depicted functions.

Example 7.3

Revisiting our Example 7.1, if $f(x)=nx^{n-1}$, then the integral
$$F(x)=\int f(x)dx=\int nx^{n-1}dx=x^n+C$$
where C is any constant.

7.2.2 Basic Indefinite Integrals

Similar to what we did for derivatives in Sect. 6.9, we here provide several basic indefinite integrals that are useful to remember. Note that when you know derivatives of functions, you actually already know a lot of indefinite integrals, as well, by thinking about the inverse operation (from the derivative back to the original function). Thus, Tables 7.1 and 6.1 bear many similarities as the derivative of the integral of a function is this function again. For example, differentiating a power function involves lowering the power by one, whereas integrating a power function involves increasing the power by one. There is an exception though, when the power is -1 (see Tables 6.1 and 7.1). Probably, the concept of indefinite integrals as anti-derivatives is becoming clearer now.

Table 7.1 Indefinite integrals $\int f(x)dx$ for basic functions $f(x)$. More basic indefinite integrals can be found at https://en. wikipedia.org/wiki/Lists_of_integrals

$f(x)$	$F(x) = \int f(x)dx$
A (constant)	$Ax + C$
x^n, $n \in \mathbb{C} \wedge n \neq -1$	$\frac{x^{n+1}}{n+1} + C$
e^{ax}, $a \in \mathbb{C} \wedge a \neq 0$	$\frac{1}{a}e^{ax} + C$
$\frac{1}{x}$ (or x^{-1})	$\begin{cases} \ln x + C & \text{if } x > 0 \\ \ln(-x) + C & \text{if } x < 0 \end{cases}$
$\sin ax$, $a \in \mathbb{C} \wedge a \neq 0$	$-\frac{1}{a}\cos ax + C$
$\cos ax$, $a \in \mathbb{C} \wedge a \neq 0$	$\frac{1}{a}\sin ax + C$
$\tan x$	$-\ln|\cos x| + C$
a^x, $a > 0 \wedge a \neq 1$	$\frac{a^x}{\ln a} + C$
$\frac{1}{\sqrt{x^2 \pm 1}}$	$\ln\left|x + \sqrt{x^2 \pm 1}\right| + C$
$\frac{1}{x^2+1}$	$\arctan x + C$
$\frac{1}{\sqrt{1-x^2}}$	$\arcsin x + C$

Example 7.4

Determine the integral of $f(x) = \frac{3}{\sqrt{5x}}$

It is easier to determine this integral once you realize that f(x) is actually a power function: $(x) = \frac{3}{\sqrt{5x}} = 3 \cdot (5x)^{-\frac{1}{2}} = 3.5^{-\frac{1}{2}}x^{-\frac{1}{2}}$. Now, we can determine the integral using Table 7.1:

$$F(x) = \int f(x)dx = \int 3 \cdot 5^{-\frac{1}{2}}x^{-\frac{1}{2}}dx = 3 \cdot 5^{-\frac{1}{2}}\int x^{-\frac{1}{2}}dx = 3 \cdot 5^{-\frac{1}{2}}\frac{x^{-\frac{1}{2}+1}}{-\frac{1}{2}+1} + C = \frac{6}{\sqrt{5}}\sqrt{x} + C$$

Before providing more examples and practicing integration yourself, we first present some basic rules of integration in Box 7.1:

Box 7.1 Basic rules of integration

1. $\frac{d}{dx}\int f(x)dx = f(x)$
2. $\int \frac{d}{dx}f(x)dx = f(x) + C$
3. $\int af(x)dx = a\int f(x)dx$, if a is a constant
4. $\int[af(x) \pm bg(x)]dx = a\int f(x)dx \pm b\int g(x)dx$, if a and b are constants (linearity).

Example 7.5

Determine the following integrals:

a. $\int \sqrt{x\sqrt{x\sqrt{x}}}dx = \int \sqrt{x\sqrt{x\sqrt{x^{\frac{1}{2}}}}}dx = \int \sqrt{x \cdot x^{\frac{3}{2} \cdot \frac{1}{2}}}dx = \int \sqrt{x^{1+\frac{3}{4}}}dx = \int x^{\frac{7}{4} \cdot \frac{1}{2}}dx =$

$\int x^{7/8}dx = \dfrac{8x^{15/8}}{15} + C$

b. $\int \left(\frac{3}{x} + \sin 5x\right)dx = \int \frac{3}{x}dx + \int \sin 5x dx = 3\ln|x| - \frac{1}{5}\cos 5x + C$

c. $\int(\sin 5x - \sin 5\alpha)dx = \int \sin 5x dx - \int \sin 5\alpha dx = -\frac{1}{5}\cos 5x - x\sin 5\alpha + C$. Note that since the integration is over x, $\sin 5\alpha$ should be considered as a constant.

d. $\int \left(e^{-i\omega t} + e^{i\omega t}\right)dt = \int e^{-i\omega t}dt + \int e^{i\omega t}dt = -\frac{1}{i\omega}e^{-i\omega t} + \frac{1}{i\omega}e^{i\omega t} + C$

Exercise

7.1 Determine the following indefinite integrals:

a. $\int(e^{3t} + 2\sin 2t)dt$
b. $\int \frac{3}{x^2}dx$
c. $\int 4x^{-1}dx$
d. $\int \sqrt{4x}dx$
e. $\int \left(\sqrt{4x^{-3}} + 5\right)dx$
f. $\int(3^x + \tan x)dx$

7.3   Definite Integrals: Integrals as Areas Under a Curve

So far, we considered integrals as anti-derivatives, thereby introducing indefinite integrals. Here, we view integrals in a different way, as areas under a curve, bounded by a lower and an upper *limit*. The curve is thus a graph of a function on a specific domain. Let's discuss this link between integrals and area under a curve in more detail. Suppose you want to know the area under the curve for the graph of the function $f(x) = -x^2 + 5$ between $x = -2$ and $x = 2$ (Fig. 7.2, left). A very rough approximation of the area under the curve would be to calculate the sum of the areas of the rectangles with a base of 1 and a height of $f(x)$ for all integer values of x from $x = -2$ to $x = 1$ (Fig. 7.2, middle). Now you can probably imagine that when we decrease the base of these rectangles by doubling the number of rectangles, the sum of their areas will better approach the area under the curve (Fig. 7.2, right). If we increase the number of rectangles even further to n and denote the base of these rectangles by Δx we find that the approximation of the area under the curve for this specific example equals

$$\sum_{i=0}^{n-1} f(-2 + i \cdot \Delta x) \cdot \Delta x$$

Now, if we let the number of rectangles between two general limits a and b (instead of -2 and 2) go to ∞ by letting Δx go to 0, we arrive at the definition of the definite integral

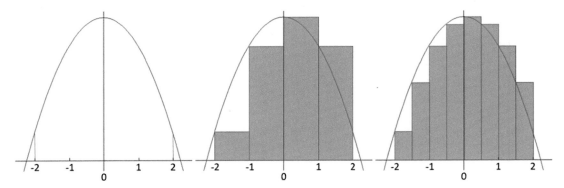

Fig. 7.2 Illustration of calculation of area under the curve for the function $f(x) = -x^2 + 5$ for $x \in (-2, 2)$. *Left*: graph of the function, *vertical lines* at $x = -2$ and $x = 2$ indicate the boundaries of the domain. *Middle*: approximation of the area under the curve when using rectangles of base 1. *Right*: improved approximation of the area under the curve when using rectangles of base 0.5.

$$\int_a^b f(x)\,dx$$

where a is the lower limit and b the upper limit. It should now be clear to you that this expression is equal to the area under the curve given by the graph of the function $f(x)$ between $x = a$ and $x = b$. More formally, the sum and integral that were used here are known as the Riemann sum and Riemann integral. Now, you may also understand that the differential dx can be understood as the limit of Δx when it goes to 0.

Also in the definition of the definite integral using the Riemann sum, we can retrace that integration is the inverse of differentiation. Remember that the formal definition of a derivative (Sect. 6.8, Eq. 6.1, replacing y by $f(x)$) was:

$$\frac{df(x)}{dx} = \lim_{\Delta x \to 0} \frac{\Delta f(x)}{\Delta x}$$

Here, we also considered small changes in x and in $f(x)$ as $\Delta f(x) = f(x + \Delta x) - f(x)$. Hence, in derivation we **subtract** function values and **divide** the difference by Δx, while in integration we **add** function values and **multiply** the sum by Δx.

To calculate the definite integral you need to determine the anti-derivative or primitive function and subtract its values at the two limits:

$$\text{If } F(x) = \int f(x)\,dx \text{ then } \int_a^b f(x)\,dx = F(b) - F(a) \tag{7.1}$$

Sometimes a slightly different notation is used:

$$\int_a^b f(x)\,dx = F(x)\big|_{x=a} - F(x)\big|_{x=b}$$

where $F(x)|_{x=a}$ should be read as $F(x)$ for $x=a$.

Example 7.6

Calculate $\int\limits_{2\pi}^{3\pi} \sin x\, dx$.

We know that (Table 7.1)

$$\int \sin x\, dx = -\cos x + C$$

By following rule (7.1), we can now calculate that:

$$\int\limits_{2\pi}^{3\pi} \sin x\, dx = (-\cos 3\pi) - (-\cos 2\pi) = -(-1) + 1 = 2$$

We now present some important rules for definite integrals in Box 7.2:

Box 7.2 Important rules for definite integrals

1. $\int\limits_a^a f(x)\, dx = 0$

2. $\int\limits_a^b f(x)\, dx = -\int\limits_b^a f(x)\, dx$

3. If $c \in (a,b)$ then $\int\limits_a^b f(x)\, dx = \int\limits_a^c f(x)\, dx + \int\limits_c^b f(x)\, dx$

These rules concern the limits of an integral. Thus, clearly any definite integrals with the same upper and lower limits are equal to zero. Swapping the upper and lower limits swaps the sign of the result. The third rule in Box 7.2 is the most interesting as it can sometimes come in quite handy when calculating definite integrals, as illustrated in the next example.

Example 7.7

Calculate

$$\int\limits_{2\pi}^{2\frac{1}{8}\pi} \sin x\, dx + \int\limits_{2\frac{1}{8}\pi}^{3\pi} \sin x\, dx$$

(continued)

Example 7.7 (continued)

Here it would be pretty hard to calculate $\cos\left(2\frac{1}{8}\pi\right)$ without a calculator, whereas when we use the third rule in Box 7.2 and employ the answer to Example 7.3, we find that

$$\int_{2\pi}^{2\frac{1}{8}\pi} \sin x\,dx + \int_{2\frac{1}{8}\pi}^{3\pi} \sin x\,dx = \int_{2\pi}^{3\pi} \sin x\,dx = 2$$

Notice that the solution to definite integrals, in contrast to indefinite integrals, does not contain a constant (C). Let's now see how definite integrals can be used to calculate the area under a complex curve (Example 7.8) and how a practical—albeit simple—problem (Example 7.9) can give us more insight in why we calculate definite integrals the way we do.

Example 7.8

Consider the function

$$f(x)=4\cos(x+0.1)-0.5x$$

in Fig. 7.3 and determine the area under its graph between the two (approximate) roots $x-1.9122$ and $x=5.3439$.

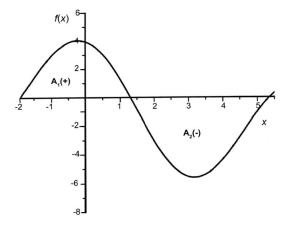

Fig. 7.3 Plot of the function $f(x)=4\cos(x+0.1)-0.5x$ for $x\in(-2,5.5)$. The area under its graph consists of a positive (A_1) and a negative (A_2) part.

Between $x=-1.9122$ and $x=5.3439$ the function is first positive, then negative. It changes sign at (approximately) $x=-1.9122$ and $x=1.3067$ and then at $x=5.3439$. Thus, the area A under the curve will be

$$A=A_1-A_2$$

(continued)

Example 7.8 (continued)

which is the so called signed area as illustrated in Fig. 7.4.

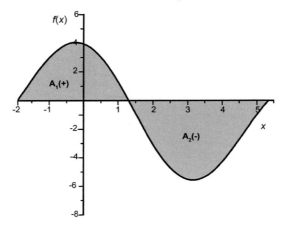

Fig. 7.4 Area under the curve (*blue*) for the function $f(x)=4\cos(x+0.1)-0.5x$ between $x=-1.9122$ and $x=5.3439$.

We now know that $A_1 = \displaystyle\int_{-1.9122}^{1.3067} f(x)dx$ and $A_2 = \displaystyle\int_{1.3067}^{5.3439} f(x)dx$. Thus, the area under the curve for $f(x)$ between $x=-1.9122$ and $x=5.3439$ is

$$A = \int_{-1.9122}^{1.3067} f(x)dx - \int_{1.3067}^{5.3439} f(x)dx$$

We can now apply the rules in Table 7.1 to find the primitive or anti-derivative F of $f(x)$ and calculate A. The primitive is $F(x) = 4\sin(x+0.1) - \frac{0.5}{2}x^2$ and thus

$$A = (F(1.3067) - F(-1.9122)) - (F(5.3439) - F(1.3067))$$
$$= 4\sin(1.3067 + 0.1)$$
$$- 4\sin(-1.9122 + 0.1) - \frac{0.5}{2}1.3067^2 + \frac{0.5}{2}(-1.9122)^2$$
$$- 4\sin(5.3439 + 0.1)$$
$$+ 4\sin(1.3067 + 0.1) + \frac{0.5}{2}5.3439^2 - \frac{0.5}{2}(1.3067)^2 \approx 21.9529$$

Example 7.9

Marianne is speed walking at a constant velocity of 2 m/s. What distance will she cover within 9 s if she keeps walking at the same speed?

We can approach this problem in two different ways. Let's first do it in a way which does not require integrals and uses our knowledge of physics. We know that distance travelled equals velocity times duration, thus Marianne covers 2 m/s × 9 s = 18 m within 9 s. A more complex way, that helps us understand why we calculate definite integrals the way we do, is the following. We know that the duration $\Delta t = 9s$. Let's assume that we determine the distance travelled between start time $t_0 = 1s$ and end time $t_{END} = 10s$. We also know that velocity is the derivative of distance travelled in time:

(continued)

Example 7.9 (continued)

$v = \frac{dx}{dt} \approx \frac{\Delta x}{\Delta t}$ or $\Delta x = v \cdot \Delta t$. Now $v \cdot \Delta t$ is the area of a rectangle with base Δt and height v which is the area under the curve of the constant function $v = 2$ m/s as displayed in the figure, or

$$x = \int_{t_0}^{t_{END}} v(t)dt = vt|_{t_0}^{t_{END}} = \frac{2m}{s}(10 - 1)s = 18m$$

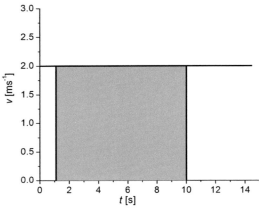

We now also see that subtracting the two values of the primitive is equal to subtracting the distance travelled in 1 s from the distance travelled in 10 s, which is the distance travelled in 9 s. Anyway, Marianne thus covers 18 m in 9 s. Is that fast enough for speed-walking?

Exercise

7.2. Determine the following definite integrals:

a. $\int_0^1 \sqrt{x^3}\,dx$

b. $\int_0^{\frac{T}{2}} \sin\left(\frac{2\pi t}{T}\right)dt$

c. $\int_0^1 (e^x - 1)^2 e^x dx$

7.3.1 Multiple Integrals

Just as we can differentiate functions of multiple variables by partial differentiation (see Sect. 6.10), we can also integrate functions of multiple variables. Such (definite) integrals are called *multiple integrals*. And like definite integrals of one variable are associated with area under a curve, definite integrals of two variables are associated with volume under a surface, defined by the domain that is integrated over. To integrate functions of multiple variables, you start from the inner most integral and work your way out, always considering the variables you are not integrating over as constant, again similar to partial differentiation, where you consider the variables you do not differentiate for as constant. Let's make this clearer by an example.

Example 7.10

Calculate $\int\limits_{1}^{2}\int\limits_{2}^{4} (xy^2 + 3x^3y)\, dx\, dy$

This integral should be read as $\int\limits_{1}^{2}\left(\int\limits_{2}^{4} xy^2 + 3x^3y\, dx\right) dy$ and thus, to calculate it, we have to integrate the function $f(x,y) = xy^2 + 3x^3y$ for x between 2 and 4 and y between 1 and 2. Executing this step-by-step, we find that:

$$\int\limits_{1}^{2}\int\limits_{2}^{4} xy^2 + 3x^3y\, dx\, dy = \int\limits_{1}^{2}\left(\frac{1}{2}x^2y^2 + \frac{3}{4}x^4y\right)\Bigg|_{2}^{4} dy = \int\limits_{1}^{2}(8y^2 + 192y) -$$

$$(2y^2 + 12y)\, dy = \int\limits_{1}^{2} 6y^2 + 180y\, dy = (2y^3 + 90y^2)\Bigg|_{1}^{2}$$

$$= (16 + 360) - (2 + 90) = 284$$

In this example, we calculated a *double integral*. Similarly, an integral with three variables of integration is called a *triple integral*.

7.4 Integration Techniques

So far, we only considered integrals of relatively simple functions. However, not all integration of relatively simple functions is simple. For example, how do we integrate functions such as $f(x) = \sqrt{x}(1 + x)$ or $f(x) = x\sin 2x$? There are numerous integration techniques that can help. Some of them are universal for all types of integrals. Some are more suited for definite integrals. Here, we will only explain some of these techniques, and illustrate them with examples. The goal of this section is to get you familiarized with the practice of integration. For more practice, we recommend starting with Jordan and Smith 2010, Mathematical techniques, and advance with Chap. 7 of Demidovich et al. (1966) available online in English.

7.4.1 Integration by Parts

Integration by parts is a method that relies on the product rule for differentiation (see also Table 6.2) that states that if

$$y(x) = f(x)g(x)$$

then

$$y'(x) = f'(x)g(x) + f(x)g'(x).$$

If we now apply anti-derivation to the first equation and substitute the second equation, we get:

$$f(x)g(x) = y(x) = \int y'(x)dx = \int (f'(x)g(x) + f(x)g'(x))dx$$

By rearranging the most outer terms, we find that:

$$\int f(x)g'(x)dx = f(x)g(x) - \int f'(x)g(x)dx \qquad (7.2)$$

This means, that if a function is the product of one function with the derivative of another function, we have a method to determine the integral of this product.

Example 7.11

Determine the integral

$$y(x) = \int x^2 e^x dx$$

To solve this integral, we first consider the simpler integral

$$y_1(x) = \int x e^x dx$$

Careful consideration of this function shows that integration by parts according to Eq. 7.2 can be applied for $f(x) = x$ and $g'(x) = e^x$. To perform integration by parts we now need to determine $f'(x)$ and $g(x)$, which is relatively simple as these are simple functions. Thus $f'(x) = 1$ and $g(x) = \int e^x dx = e^x$ and after integration by parts of y_1 we find that:

$$y_1(x) = \int x e^x dx = x e^x - \int 1 \cdot e^x dx = x e^x - e^x + C = e^x(x - 1) + C$$

Simple, right? This is how integration by parts works. Now let's go back to the original function $y(x)$. For this function, we choose $f(x) = x^2$ and $g'(x) = e^x$ and thus $f'(x) = 2x$ and $g(x) = e^x$. Applying integration by parts according to Eq. 7.2 we now find:

$$y(x) = \int x^2 e^x dx = x^2 e^x - \int 2x e^x dx$$

We have just determined that $\int x e^x dx = e^x(x - 1) + C$ and thus:

$$y(x) = x^2 e^x - 2(e^x(x - 1) + C) = e^x(x^2 - 2x + 2) + C$$

For integration by parts to be applicable you should be able to assign $f(x)$ and $g'(x)$ such that

1. $f'(x)$ is simpler than $f(x)$
2. $g(x)$ is not more complicated than $g'(x)$

Example 7.12

Determine $\int x \ln x \, dx$.

Here we consider the two functions x and $\ln x$ that form the product. Is one of them simpler after derivation? Is the second one not more complicated after integration? The derivative of $f(x) = \ln x$ is in fact simpler than the function itself:

$$f'(x) = \frac{1}{x}$$

Thus, we have identified one function that is simpler after differentiation. But does the second function in the product have an integral that is not more complicated than the function itself? So, let's assign $g'(x) = x$. Then $g(x) = \int x \, dx = \frac{x^2}{2}$. This integral is not much more complicated, but is it sufficiently simple? Let's try and find out:

Applying integration by parts according to Eq. 7.2, using our choice of $f(x)$ and $g'(x)$ we find that:

$$\int x \ln x \, dx = \frac{x^2}{2} \ln x - \int \frac{1}{x} \frac{x^2}{2} \, dx = \frac{x^2}{2} \ln x - \frac{x^2}{4} + C = \frac{x^2}{2} \left(\ln x - \frac{1}{2} \right) + C$$

Example 7.13

Determine $\int e^x \cos x \, dx$.

Now we consider the two functions e^x and $\cos x$ that form the product. Is one of them simpler after derivation? Is the second one not more complicated after integration?

We know that e^x is not more complicated after integration. But is $\cos x$ simpler after derivation? Let's give it a try and assign $f(x) = \cos x$, $g'(x) = e^x$. Then $f'(x) = -\sin x$ and $g(x) = e^x$. When we now apply integration by parts according to Eq. 7.2 we find that:

$$\int e^x \cos x \, dx = \cos x e^x + \int \sin x e^x \, dx$$

The integral on the right-hand side actually does not look much simpler. But let's keep at it and apply integration by parts one more time to the integral on the right-hand side. This time we assign $f(x) = \sin x$, $g'(x) = e^x$. Then $f'(x) = \cos x$ and $g(x) = e^x$ and we find that:

$$\int e^x \cos x \, dx = \cos x e^x + \sin x e^x - \int e^x \cos x \, dx$$

Now we recognise the integral that we want to determine ($\int e^x \cos x \, dx$) on both sides of the equation. By rearranging the terms, we get:

$$2 \int e^x \cos x \, dx = \cos x e^x + \sin x e^x + C$$

and thus:

$$\int e^x \cos x \, dx = \frac{e^x}{2} (\cos x + \sin x) + C$$

Exercise

7.3 Determine the following integrals using integration by parts:

a. $\int (x-1)^2 e^x dx$

b. $\int_0^\pi x \sin 2x dx$

c. $\int (\ln x)^2 dx$

7.4.2 Integration by Substitution

Another, probably much more often used method to determine integrals is $(u\text{-})$substitution. In short, the aim is to make the integrand as simple as possible to determine the integral. To determine integrals by rewriting them to "easier" forms the following steps need to be taken:

1. For an integral $\int f(x) dx$ find a part of the function that can be substituted by $u(x)$
2. Differentiate to express dx in terms of du: $dx = x'(u) du$
   Check if the new integral $\int f(x) dx = \int f(x(u)) x'(u) dx$ is easier to solve and if not try another substitution.
3. Once you have finished the calculus, substitute back to the initial variable x to find an indefinite integral or also substitute the limits to find a definite integral.

As this explanation probably sounds rather abstract, let's try to get a better understanding by some examples.

Example 7.14

Determine $\int x(x+5)^5 dx$.

First, we determine a likely suitable substitution $u(x)$. It seems appropriate to simplify the function by substituting

$$u(x) = x+5$$

We can find the differential du as follows:

$$du = d(x+5) = dx + 0 = dx$$

We also know that:

$$x = u - 5$$

Now we can rewrite the integral as follows:

$$\int x(x+5)^5 dx = \int (u-5)(u)^5 du$$

This is easy to solve for u:

(continued)

Example 7.14 (continued)

$$\int (u - 5)(u)^5 du = \int (u^6 - 5u^5) du = \int u^6 du - 5 \int u^5 du = \frac{u^7}{7} - \frac{5u^6}{6} + C$$

To determine the original integral in terms of x, all we should do now is substitute u again. Thus, the integral is:

$$\int x(x + 5)^5 dx = \frac{(x + 5)^7}{7} - \frac{5(x + 5)^6}{6} + C$$

Example 7.15

Determine $\int (3x^2 + 10x + 7) e^{-(5x^2 + 7x + x^3)} dx$

At first sight, the integrand looks too complex to be able to determine the integral. But let's try to find a suitable substitute $u(x)$. As the biggest problem seems to be in the complex exponent, we first try to define this as a substitute:

$$u(x) = 5x^2 + 7x + x^3$$

As we know that $\int e^{-x} dx = -e^{-x} + C$, this might be a sensible approach to this integral. Next, we determine the differential du by derivation:

$$\frac{du}{dx} = 10x + 7 + 3x^2$$

or

$$du = (3x^2 + 10x + 7) dx$$

Now substitution results in a very simple integrand and integration becomes a piece of cake:

$$\int (3x^2 + 10x + 7) e^{-(5x^2 + 7x + x^3)} dx = \int e^{-u} du = -e^{-u} + C$$

Finally, substituting $u(x)$ results in:

$$\int (3x^2 + 10x + 7) e^{-(5x^2 + 7x + x^3)} dx = -e^{-(5x^2 + 7x + x^3)} + C$$

Example 7.16

Determine $\int \sqrt{5x + 3} dx$.

The most obvious choice for substitution is:

$$u(x) = 5x + 3$$

(continued)

Example 7.16 (continued)

Then

$$\frac{du(x)}{dx} = 5$$

$$dx = \frac{1}{5}du$$

We can now rewrite the integral to:

$$\int \sqrt{5x + 3}\,dx = \int \sqrt{u}\frac{1}{5}\,du = \frac{1}{5}\frac{2}{3}u^{\frac{3}{2}} + C$$

For the final step, substitution gives:

$$\int \sqrt{5x + 3}\,dx = \frac{2}{15}\sqrt{(5x + 3)^3} + C$$

Example 7.17

Determine $\int \frac{3x^4}{x^5+6}\,dx$.

There is no obvious choice for substitution now: we could either choose the numerator or the denominator as a candidate for substitution. However, the denominator has a higher order polynomial than the numerator. Thus, if the denominator is differentiated it will be closer to the numerator. For this reason, we choose the denominator for u- substitution:

$$u(x) = x^5 + 6$$

$$\frac{du(x)}{dx} = 5x^4$$

$$dx = \frac{1}{5x^4}du$$

We can now rewrite the integral to:

$$\int \frac{3x^4}{x^5 + 6}\,dx = \int \frac{3x^4}{u}\frac{1}{5x^4}\,du = \frac{3}{5}\int \frac{du}{u} = \frac{3}{5}\ln|u| + C$$

So finally, substituting u again yields:

$$\int \frac{3x^4}{x^5 + 6}\,dx = \frac{3}{5}\ln|x^5 + 6| + C$$

Example 7.18

Determine $\int \cos^6 x \sin x dx$.

   If we again choose to substitute the part of the product with the higher power, similar to Example 7.16, we can write:

$$u(x) = \cos x$$

$$\frac{du(x)}{dx} = -\sin x$$

$$dx = \frac{-1}{\sin x} du$$

We can now rewrite the integral to:

$$\int \cos^6 x \sin x dx = \int u^6 \sin x \frac{-1}{\sin x} du = -\int u^6 du = -\frac{1}{7} u^7 + C$$

Final substitution of u gives:

$$\int \cos^6 x \sin x dx = -\frac{1}{7} (\cos x)^7 + C$$

Exercise

7.4 Determine the following integrals using substitution:

   a. $\int x^2 e^{-4x^3} dx$

   b. $\int \frac{3 \sin x}{2 + \cos x} dx$

   c. $\int \frac{(\sqrt{x}+2)^6}{\sqrt{x}} dx$

   d. $\int \frac{3}{x \ln x} dx$

   e. $\int_0^1 \sqrt{1 + x} dx$

7.4.3 Integration by the Reverse Chain Rule

Just like integration by parts employed the product rule for differentiation, we can use the chain rule for differentiation to our advantage for integration. One could say that the "reverse chain rule" makes implementation of u-substitution easier as, in a way, it is the same rule. Remember that for a composite function $f(x) = g(h(x))$ (Table 6.2):

$$\frac{df(x)}{dx} = \frac{dg(h(x))}{dx} = \frac{dg}{dh}\frac{dh}{dx} = h'(x)g'(h(x))$$

Applying integration, we find that:

$$\int h'(x)g'(h(x))dx = g(h(x)) + C \qquad (7.3)$$

In general, this rule is used for integration of trigonometric, logarithmic, rational/power and exponential functions. To apply the reverse chain rule in case of some composite function (e.g. $\sin(3x+5)$, or $\log_5|\sin x|$), one should try to recognise the derivative of the function inside the composite function (thus $h(x)$). For example, let's consider the integrand:

$$x\sin x^2$$

Note that $\sin x^2 = \sin(x^2)$, whereas $\sin^2 x = \sin x \sin x$. In this case $2x$ is the derivative of x^2 and we recognize that the sinusoidal function is multiplied by half the derivative of the function that is inside the sinusoidal function. We can thus write:

$$x \sin x^2 = 2x \cdot \tfrac{1}{2} \sin x^2 = h'(x)g'(h(x))$$

Once we have recognized this structure in the integrand, the next step is to determine $h(x)$ and $g(h)$:

$$h(x) = x^2, \qquad h'(x) = 2x$$

$$g'(h) = \frac{1}{2} \sin h, \qquad g(h) = -\frac{1}{2} \cos h$$

We then know how to determine the integral of this composite function, by applying the reverse chain rule Eq. 7.3:

$$\int x \sin x^2 dx = -\tfrac{1}{2} \cos x^2 + C$$

Example 7.19

Determine $\int 3x^2 \sin x^3 dx$.
   We recognize that $3x^2$ is the derivative of x^3 and hence apply the reverse chain rule (Eq. 7.3) as follows:

$$h(x) = x^3, \qquad h'(x) = 3x^2,$$
$$g'(h) = \sin h, \qquad g(h) = -\cos h$$

We thus find that:

$$\int 3x^2 \sin x^3 dx = -\cos x^3 + C$$

Exercise

7.5 Determine the following integrals using the reverse chain rule:

  a. $\int x^3 e^{x^4} dx$
  b. $\int x^3 (1+x^4)^3 dx$

7.4.4 Integration of Trigonometric Functions

Although the integration techniques introduced in Sects. 7.4.1 to 7.4.3 allow integration of many different functions, there will still be integrals left that cannot be determined. For specific types of integrands, such as rational or transcendental functions, several additional useful methods exist to determine their integrals (see for example Bronshtein et al. 2007). Here, we will only briefly cover some integration methods for trigonometric integrands so that you get a feeling of how integration is done in general. We direct you to more specialized literature for broader and more advanced methods of integration (e.g. Jordan and Smith (2010)). An overview can also be found on https://en.wikipedia.org/wiki/List_of_integrals_of_trigonometric_functions.

The following methods can be used to determine some integrals of trigonometric functions:

1. If the integral contains a rational function of sines and cosines, the following substitution is often useful:
   $\tan \frac{x}{2} = t$ which implies that $\sin x = \frac{2t}{1+t^2}$, $\cos x = \frac{1-t^2}{1+t^2}$ and $dx = \frac{2dt}{1+t^2}$. Here we use that $\sin x = 2 \sin \frac{x}{2} \cos \frac{x}{2}$ and $\cos x = \cos^2 \frac{x}{2} - \sin^2 \frac{x}{2}$. By this substitution, the integrand becomes a rational function of t.
2. If the integrand is a positive power of a trigonometric function, recurrent formulas can be used to determine the integral:

$$\int \sin^n x dx = \frac{1}{n} \sin^{n-1} x \cos x + \frac{n-1}{n} \int \sin^{n-2} x dx$$

$$\int \cos^n x dx = \frac{1}{n} \cos^{n-1} x \sin x + \frac{n-1}{n} \int \cos^{n-2} x dx$$

3. If the integrand is a negative power of a trigonometric function, such as $\int \frac{1}{\sin^n x} dx$ or $\int \frac{1}{\cos^n} x dx$ $(n \in N, n > 1)$, the integral can be determined by using the following recurrent formulas:

$$\int \frac{1}{\sin^n x} dx = \frac{1}{n-1} \frac{\cos x}{\sin^{n-1} x} + \frac{n-2}{n-1} \int \frac{1}{\sin^{n-2} x} dx$$

$$\int \frac{1}{\cos^n x} dx = \frac{1}{n-1} \frac{\sin x}{\cos^{n-1} x} + \frac{n-2}{n-1} \int \frac{1}{\cos^{n-2} x} dx$$

4. Integrals of the form $\int \sin^m x \cos^n x\, dx$, $(n, m \in Z, >0)$, can be determined using the following recurrent formulas:

$$\int \sin^n x \cos^m x\, dx = \frac{\sin^{n+1} x \cos^{m-1} x}{m+n} + \frac{m-1}{m+n} \int \sin^n x \cos^{m-2} x\, dx$$

or

$$\int \sin^n x \cos^m x\, dx = -\frac{\sin^{n-1} x \cos^{m+1} x}{m+n} + \frac{n-1}{n+m} \int \sin^{n-2} x \cos^m x\, dx$$

(see also https://en.wikipedia.org/wiki/List_of_integrals_of_trigonometric_functions#Integra nds_involving_both_sine_and_cosine)

5. Integrals of the form $\int \sin nx \cos mx\, dx$, $(n, m \in Z)$ can be simplified and subsequently determined using the trigonometric identities for multiplication of trigonometric functions:

$$\sin nx \cos mx = \frac{1}{2}\left[\sin(n-m)x + \sin(n+m)x \right]$$

Example 7.20

Determine $\int \frac{dx}{5+4\cos x}$

This integral can be determined using the first method in this section by substituting $\tan \frac{x}{2} = t$, $\cos x = \frac{1-t^2}{1+t^2}$ and $dx = \frac{2dt}{1+t^2}$:

$$\int \frac{dx}{5+4\cos x} = \int \frac{\frac{2dt}{1+t^2}}{5+4\frac{1-t^2}{1+t^2}} = \int \frac{2dt}{9+t^2}$$

We know how to determine this integral as it is similar to the integral of $\frac{1}{x^2+1}$ (Table 7.1)

$$\int \frac{dx}{5+4\cos x} = \frac{2}{9} \int \frac{dt}{1+\frac{t^2}{9}}$$

If we substitute $\frac{t^2}{9} = u^2$, then $dt = 3du$ and thus

---

(continued)

Example 7.20 (continued)

$$\int \frac{dx}{5+4cosx} = \frac{6}{9}\int \frac{du}{1+u^2} = \frac{6}{9}\arctan u + C = \frac{6}{9}\arctan\frac{t}{3} + C = \frac{6}{9}\arctan\tan\frac{\frac{x}{2}}{3} + C$$

7.5 Scientific Examples

7.5.1 Expected Value

The expected value $\langle x \rangle$ of a *stochastic variable* x is frequently encountered in statistics. It refers to the value one expects to get for x on average if an experiment would be run many times. For example, if you toss a coin 10 times, you expect to get 5 heads and 5 tails. You expect this value because the probability of getting heads is 0.5 and if you toss 10 times you anticipate that you will get heads 5 times. The expected value is also called the expectation value, the mean, the mean value, the mathematical expectation and, in statistics, it is known as the first moment.

If the probability distribution $P(x)$, which describes the probability of getting a specific value of x is known, the expected value can be calculated by multiplying each of the possible outcomes by the probability that each outcome will occur, and by integrating all products:

$$a = \langle x \rangle = \int xP(x)dx$$

The expected value can be viewed as the weighted average value, where the weight is given by the probability distribution. This is easier to understand in the case of a discrete distribution. Note that nowadays almost all measurements are discrete, even when measuring continuous events, as our digital devices sample the values at a specific frequency, e.g. at 2 Hz (every 0.5 s). As explained in Sect. 7.3, in the case of n discrete values, we can replace the integral by a sum

$$a = \sum_{i=1}^{i=n} x_i P(x_i)$$

Example 7.21

The most frequently used example of expected value concerns throwing a dice. If a dice is of good quality, the probability of the dice landing on any of the 6 sides is equal. Thus, the probability of getting a 5 is 1/6, as is the probability of getting any of the other possible values:

$$P(x_i) = \frac{1}{6}, \quad x_i = \{1, 2, 3, 4, 5, 6\}$$

The expected value when throwing a dice is thus:

(continued)

Example 7.21 (continued)

$$a = \sum_{i=1}^{n} x_i P(x_i) = \frac{(1 + 2 + 3 + 4 + 5 + 6)}{6} = 3.5$$

This is the same as the average of the values on a six-sided dice.

In most cases the probability will not be equal for all values of x. To calculate the expected value, it is then convenient to have a functional description of the probability distribution.

Example 7.22

A frequently encountered distribution is the *Gaussian distribution*. Any stochastic variable that is determined by many independent factors will follow such a distribution. Examples are height and weight of humans. The Gaussian probability distribution function is:

$$G(x) = \frac{1}{\sqrt{2\pi}\sigma} e^{\frac{(x-\mu)^2}{2\sigma^2}}$$

Here, μ is the mean value of the distribution and σ is its standard deviation. Let's see if μ is indeed the same as the expected value:

$$\langle x \rangle = \int_{-\infty}^{\infty} x P(x)\,dx = \int_{-\infty}^{\infty} x \frac{1}{\sqrt{2\pi}\sigma} e^{\frac{(x-\mu)^2}{2\sigma^2}}\,dx$$

To determine this integral is beyond the scope of this book, but we will here describe some important steps. For a more detailed explanation we refer to Reif (1965, the Berkley Physics course Volume 5, Appendix 1; https://en.wikipedia.org/wiki/List_of_integrals_of_exponential_functions).

The integral can be solved using substitution and then incorporating some of the characteristics of the Gaussian distribution. In this manner, the expected value can be rewritten using:

$$u = x - \mu$$
$$du = dx$$

resulting in:

$$\langle x \rangle = \frac{1}{\sqrt{2\pi}\sigma} \int_{-\infty}^{\infty} u e^{\frac{u^2}{2\sigma^2}}\,du + \frac{1}{\sqrt{2\pi}\sigma} \int_{-\infty}^{\infty} \mu e^{\frac{u^2}{2\sigma^2}}\,du$$

The first integral vanishes as the integrand is odd:

$$\int_{-\infty}^{0} u e^{\frac{u^2}{2\sigma^2}}\,du = -\int_{0}^{\infty} u e^{\frac{u^2}{2\sigma^2}}\,du$$

The second integral can be rewritten as (Reif 1965, Berkley Physics course, Volume 5, Appendix 1, Equations 11 and 12)

$$\int_{-\infty}^{\infty} \mu e^{\frac{u^2}{2\sigma^2}}\,du = \mu \int_{-\infty}^{\infty} e^{\frac{u^2}{2\sigma^2}}\,du = \mu\sqrt{2\pi}\sigma$$

Thus, the expected value of a Gaussian distribution is indeed equal to its mean.

7.5.2 Convolution

Convolution is a very important mathematical operation in the analysis of time series. It can be viewed as a type of filter. If you have one function or time series $f(t)$, convolution with another function $g(t)$ yields the amount by which $g(t)$ overlaps with $f(t)$ when $g(t)$ is shifted in time. Convolution is expressed by an integral, as follows:

$$[f * g](t) = \int_{-\infty}^{\infty} f(\tau)g(t - \tau)d\tau$$

In other words, convolution is a mathematical operation on two functions, resulting in a third function that represents the overlap between the two functions as a function of the translation of one of the original functions with respect to the other. The effect of convolution will become clearer in some examples. Note that in all figures belonging to the examples below both the convolution and $f(t)$ were normalized to the maximum of $g(t)$, for illustration purposes.

Example 7.23

The convolution of a rectangular function and a linear function results in a saw-tooth function. When the rectangular function is shifted, the maximum of the convolution, that indicates where both functions have maximum overlap, also shifts (Fig. 7.5).

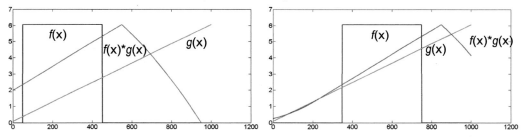

Fig. 7.5 *Left*: Convolution (*red*) of a rectangular function (*blue*) with a linear function (*green*). *Right*: Convolution of the same functions when $f(x)$ is shifted along the x-axis. Note that the convolution and $f(x)$ were normalized for illustration purposes.

Example 7.24

The convolution of a rectangular and a saw-tooth function is similar to Example 7.23. If the order of the convolution between saw-tooth and rectangular function is reversed, we see that the resulting function is shifted along the x-axis (Fig. 7.6).

(continued)

Example 7.24 (continued)

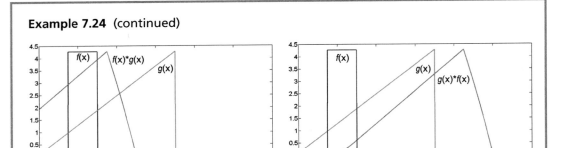

Fig. 7.6 *Left*: Convolution (*red*) of a rectangular function (*blue*) with a saw-tooth function (*green*). *Right*: Convolution of the same functions in reverse order. Note that the convolution and *f(x)* were normalized for illustration purposes.

Example 7.25

The convolution of a rectangular and a Gaussian function is again a Gaussian function (Fig. 7.7).

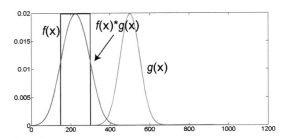

Fig. 7.7   Convolution (*red*) of a rectangular function (*blue*) with a Gaussian function (*green*).

Example 7.26

It is not true that the convolution of any function with a Gaussian function is again a Gaussian function, as is illustrated with this example of the convolution of a saw-tooth function with a Gaussian function (Fig. 7.8).

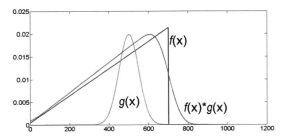

Fig. 7.8   Convolution (*red*) of a saw-tooth function (*blue*) with a Gaussian function (*green*). The convolution is skewed just like the saw-tooth function.

Example 7.27

In behavioral neuroscience, *functional magnetic resonance imaging* (*fMRI*) is often used to determine brain activation during tasks. It employs the local relative increase in oxygenated blood (the blood oxygen-level-dependent or BOLD-response) that develops when a brain area is involved in such a task. The *general linear modeling* (GLM) approach to analyze fMRI data was introduced in Chap. 5, and the BOLD response and *hemodynamic response function* (HRF) were introduced in Chap. 6 (Sect. 6.13.3). General linear modeling involves modeling the BOLD response in a brain area that is involved in a task. The simplest model assumes that there is (constant) activity in such a brain area during the task and no activity during rest, in between task blocks, resulting in a so-called block model (see Fig. 4.16, top and cf. Example 4.4). Such a model does not take the sluggish BOLD-response into account, however, which becomes maximal only seconds after the brain has been stimulated. To compensate for this sluggishness and model the specific physiological response as well as possible the block model is convoluted with a model of the BOLD response, the HRF (see Fig. 7.9 middle and bottom).

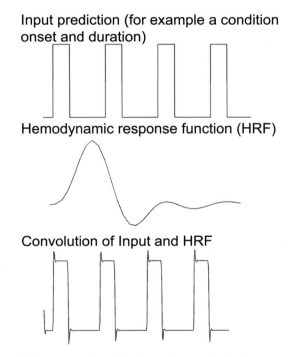

Input prediction (for example a condition onset and duration)

Hemodynamic response function (HRF)

Convolution of Input and HRF

Fig. 7.9 fMRI data analysis preparation. The input time series (*top*) are convoluted with the HRF (*middle*) to create a predictor for the GLM analysis (*bottom*). This is done for each condition, task or any other experimental manipulation.

Example 7.28

Since a few decades, we mostly only make digital photos, which allows easy manipulation using computer programs such as Photoshop or Paint. Most of us have used these programs to beautify ourselves and we have become quite used to 'photoshopped' pictures of celebrities in magazines. Oftentimes, convolution is used to enhance photos digitally. For example, the

(continued)

Example 7.28 (continued)

pixelated photo on the left in Fig. 7.10 can be convoluted with a 2D Gaussian function to blur or smoothen it, so that the pixels (and thereby the wrinkles!) are not recognizable any more (Fig. 7.10, right).

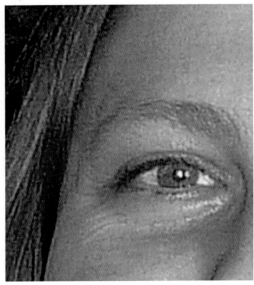 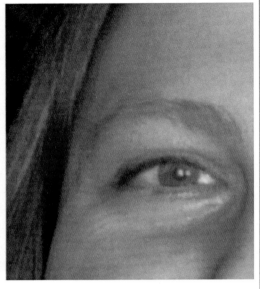

Fig. 7.10 *Left*: A pixelated picture. *Right*: The same picture after 2D Gaussian smoothing (over 4 × 4 pixels) has been applied. Wrinkles have disappeared and facial features have changed.

Example 7.29

Cross-correlation provides a beautiful example of convolution in everyday science. Cross-correlation provides a measure of the similarity between two functions as a function of a time-lag applied to one of them. This is also known as the sliding dot product or sliding inner-product (cf. Sect. 4.2.2.1). For continuous functions $f(t)$ and $g(t)$ the cross-correlation is given by

$$f * g(\tau) = \int\limits_{-\infty}^{\infty} f^*(t)g(t+\tau)dt$$

Here, $f^*(t)$ denotes the complex conjugate of $f(t)$. You can see that cross-correlation is the same as convolution for $f(-t)$. Thus, the cross-correlation is a function of τ, which has the same range as t.

Glossary

Analytic As in analytic expression, a mathematical expression that is written such that it can easily be calculated. Typically, it contains the basic arithmetic operations (addition, subtraction, multiplication, division) and operators such as exponents, logarithms and trigonometric functions.

Convolution Convolution of a function or time series $f(t)$, with another function $g(t)$ yields the amount by which $g(t)$ overlaps with $f(t)$ when $g(t)$ is shifted in time; convolution can be viewed as a modifying function or filter.

Cross-correlation A measure of similarity of two functions as a function of the displacement of one with respect to the other.

Definite As in definite integral; the integral of a function on a limited domain.

Double integral Multiple integral with two variables of integration.

fMRI Functional magnetic resonance imaging; a neuroimaging technique that employs magnetic fields and radiofrequency waves to take images of e.g. the functioning brain, employing that brain functioning is associated with changes in oxygenated blood flow.

Gaussian distribution Also known as normal distribution. It is a symmetric, bell-shaped distribution that is very common and occurs when a stochastic variable is determined by many independent factors.

General linear model Multiple linear regression; predicting a dependent variable from a set of independent variables according to a linear model.

Hemodynamic response function A model function of the increase in blood flow to active brain neuronal tissue.

Indefinite As in indefinite integral; the integral of a function without specification of a domain.

Integrand Function that is integrated.

Inverse As in 'inverse operation' or 'inverse function', meaning the operation or function that achieves the opposite effect of the original operation or function. For example, integration is the inverse operation of differentiation and $\ln x$ and e^x are each other's inverse functions.

Limit Here: the boundaries of the domain for which the definite integral is determined.

Multiple integral Definite integral over multiple variables.

Numerical integration Estimating the value of a definite integral using computer algorithms.

Primitive Anti-derivative.

Stochastic variable A variable whose value depends on an outcome, for example the result of a coin toss, or of throwing a dice.

Triple integral Multiple integral with three variables of integration.

Symbols Used in This Chapter (in Order of Their Appearance)

$\int \cdot$	Indefinite integral
$\int_a^b \cdot$	Definite integral between the limits a and b
arctan	Inverse of tangent function
arcsin	Inverse of sine function
$\approx$	Approximately equal to

Overview of Equations for Easy Reference

Indefinite integral

$$f(x) = \int f'(x)\,dx$$

Basic indefinite integrals

$$f(x) = A \text{ (constant)}, \qquad \int f(x)dx = Ax + C$$

$$f(x) = x^n, \quad n \in \mathbb{C} \wedge n \neq -1, \quad \int f(x)dx = \frac{x^{n+1}}{n+1} + C$$

$$f(x) = e^{ax}, \quad a \in \mathbb{C} \wedge a \neq 0 \qquad \int f(x)dx = \frac{1}{a}e^{ax} + C$$

$$f(x) = \frac{1}{x} \text{ (or } x^{-1}) \qquad \int f(x)dx = \begin{cases} lnx + C & \text{if } x > 0 \\ \ln(-x) + C & \text{if } x < 0 \end{cases}$$

$$f(x) = \sin ax, \quad a \in \mathbb{C} \wedge a \neq 0, \quad \int f(x)dx = -\frac{1}{a}\cos ax + C$$

$$f(x) = \cos ax, \quad a \in \mathbb{C} \wedge a \neq 0 \qquad \int f(x)dx = \frac{1}{a}\sin ax + C$$

$$f(x) = \tan x, \qquad \int f(x)dx = -\ln|\cos x| + C$$

$$f(x) = a^x, \quad a > 0 \wedge a \neq 1, \qquad \int f(x)dx = \frac{a^x}{\ln a} + C$$

$$f(x) = \frac{1}{\sqrt{x^2 \pm 1}}, \qquad \int f(x)dx = \ln\left|x + \sqrt{x^2 \pm 1}\right| + C$$

$$f(x) = \frac{1}{x^2 + 1}, \qquad \int f(x)dx = \arctan x + C$$

$$f(x) = \frac{1}{\sqrt{1 - x^2}}, \qquad \int f(x)dx = \arcsin x + C$$

Basic rules of integration

1. $\frac{d}{dx}\int f(x)dx = f(x)$
2. $\int \frac{d}{dx}f(x)dx = f(x) + C$
3. $\int af(x)dx = a\int f(x)dx$, if a is a constant
4. $\int [af(x) \pm bg(x)]dx = a\int f(x)dx \pm b\int g(x)dx$, if a and b are constants (linearity).

Definite integral

$$\int_a^b f(x)dx$$

where a and b are the limits of integration.

If $F(x) = \int f(x)dx$ then $\int_a^b f(x)dx = F(b) - F(a)$ or $\int_a^b f(x)dx = F(x)|_{x=a} - F(x)|_{x=b}$

where $F(x)|_{x=a}$ is $F(x)$ for $x=a$.

Important rules for definite integrals

1. $\displaystyle\int_{a}^{a} f(x)\,dx = 0$

2. $\displaystyle\int_{a}^{b} f(x)\,dx = -\int_{b}^{a} f(x)\,dx$

3. If $c \in (a,b)$ then $\displaystyle\int_{a}^{b} f(x)\,dx = \int_{a}^{c} f(x)\,dx + \int_{c}^{b} f(x)\,dx$

Integration by parts

$$\int f(x)g'(x)\,dx = f(x)g(x) - \int f'(x)g(x)\,dx$$

Reverse chain rule

$$\int h'(x)g'(h(x))\,dx = g(h(x)) + C$$

Expected value

$$a = \langle x \rangle = \int x P(x)\,dx$$

Convolution

$$[f * g](t) = \int_{-\infty}^{\infty} f(\tau)g(t - \tau)\,d\tau$$

Cross-correlation

$$f * g(\tau) = \int_{-\infty}^{\infty} f^*(t)g(t + \tau)\,dt$$

Answers to Exercises

7.1. a. $\frac{1}{3}e^{3t} - \cos 2t + C$

    b. $\frac{-3}{x} + C$

    c. $4\ln|x| + C$

    d. $\frac{4}{3}x^{3/2} + C$

    e. $-4x^{-\frac{1}{2}} + 5x + C$

    f. $\frac{3^x}{\ln 3} - \ln|\cos x| + C$

7.2. a. $\frac{2}{5}$

b. $\frac{T}{\pi}$

c. $e\left(\frac{1}{3}e^2 - e + 1\right) - \frac{1}{3}$

7.3. a. $(x-1)^2 e^x - 2e^x(x-1) + 2e^x + C = e^x(x^2 - 4x + 5) + C$

b. $-\frac{\pi}{2}$

c. In this case, we suggest to use integration by parts twice. First, we write

$f(x) = (\ln x)^2$, $f'(x) = 2\frac{1}{x}\ln x$. That leaves us with $g'(x) = 1$, thus $g(x) = x$.
So, applying integration by parts once, we find that:

$$\int (\ln x)^2 dx = x(\ln x)^2 - \int \frac{2x}{x} \ln x dx = x(\ln x)^2 - 2 \int \ln x dx$$

The remaining integral on the right-hand side, we can solve by again applying integration by parts. This time we choose $f(x) = \ln x$, $f'(x) = \frac{1}{x}$, and as above $g'(x) = 1$, thus $g(x) = x$.

$$\int \ln x dx = x \ln x - \int dx = x \ln x - x + C$$

So, we finally arrive at:

$$\int (\ln x)^2 dx = x(\ln x)^2 - 2x \ln x + 2x + C$$

7.4. a. $-\frac{1}{12}e^{-4x^3} + C$

b. $-3\ln|2 + \cos x| + C$

c. $\frac{2}{7}(\sqrt{x} + 2)^7 + C$

d. $3\ln|\ln|x|| + C$

e. $\frac{2}{3}(\sqrt{8} - 1)$

7.5. a. Use

$$h(x) = x^4, \qquad h'(x) = 4x^3,$$
$$g'(h) = e^h, \qquad g(h) = e^h$$

The result is $\frac{1}{4}e^{x^4} + C$

b. Use

$$h(x) = \frac{1}{4}x^4, \qquad h'(x) = x^3,$$
$$g'(h) = (1 + 4h)^3, \qquad g(h) = \frac{1}{16}(1 + 4h)^4$$

The result is $\frac{1}{16}\left(1 + x^4\right)^4 + C$

References

Online Sources of Information

https://www.youtube.com/watch?v=EUDKcjWG1ck
https://www.youtube.com/watch?v=Ma0YONjMZLI
https://en.wikipedia.org/wiki/Lists_of_integrals
https://en.wikipedia.org/wiki/Riemann_integral
https://en.wikipedia.org/wiki/List_of_integrals_of_trigonometric_functions#Integrands_involving_both_sine_and_cosine

Books

Demidovich B.P. 1966, Problems in Mathematical Analysis. MIR, Moscow , https://archive.org/details/problemsinmathem031405mbp

D.W. Jordan, P. Smith, *Mathematical Techniques*, 4th edn. (Oxford University Press, New York, 2010)

F. Reif, *Berkley Physics Course, Volume 5* (McGraw-Hill, Berkeley, 1965)

I.N. Bronshtein, K.A. Semendyayev, G. Musiol, H. Mühlig, *Handbook of Mathematics* (Springer, New York, 2007)

Index

© Springer International Publishing AG 2017
N. Maurits, B. Ćurčić-Blake, *Math for Scientists*, DOI 10.1007/978-3-319-57354-0

Made in the USA
San Bernardino, CA
15 July 2018

82448119R00139